ALLE ZEIT WACH
1842

Johannes Dietl (Ed.)

The Mammalian Egg Coat

Structure and Function

Foreword by
P. M. Wassarman

With 66 Figures in 117 Separate Illustrations

Springer-Verlag Berlin Heidelberg New York
London Paris Tokyo

Priv.-Doz. Dr. med. Johannes Dietl
Universitäts-Frauenklinik
Eberhard-Karls-Universität
Schleichstraße 4
D-7400 Tübingen
Federal Republic of Germany

ISBN-13:978-3-642-74050-3 e-ISBN-13:978-3-642-74048-0
DOI: 10.1007/978-3-642-74048-0

Library of Congress Cataloging-in-Publication Data
The Mammalian egg coat. 1. Zona pellucida. 2. Mammals--Eggs. I. Dietl, Johannes, 1948- QL965.M33 1989 599'.033 88-34895
ISBN-13:978-3-642-74050-3 (U.S.)

Softcover reprint of the hardcover 1st edition 1989

2123/3145-543210 - Printed on acid-free paper

To Anna Katharina

Foreword

Publication of this monograph acknowledges great strides made in research on the zona pellucida during the past decade or so. The body of work presented testifies to a resurgence of interest in the zona pellucida, prompted largely by a renewed appreciation of its unique properties and its important functions during early mammalian development.

It is gratifying to realize that the zona pellucida has at last achieved a well-deserved attention. Whereas extracellular coats of nonmammalian eggs have been studied intensively for many years, the zona pellucida by comparison, has until now taken a back seat. Fortunately, the research presented here signals an end to this situation. Many of the difficulties inherent in working with relatively small amounts of material have been overcome, paving the way for application of a wider variety of experimental approaches to the zona pellucida by a larger community of investigators. Hopefully, this monograph will serve as a catalyst in this regard.

As revealed in these pages, application of contemporary experimental methodologies has had significant impact on both pure and applied research on the zona pellucida. For example, modern immunological approaches have been used to assess molecular features of zona pellucida structure and function, as well as to evaluate the zona pellucida as a potential contraceptive target antigen. Recombinant DNA technology, used thus far to clone zona pellucida genes and study their expression, continues to hold much promise for future work on reproductive failure attributable to zona pellucida dysfunction. Improved in vitro cell culture and biochemical fractionation methodologies have been used to dissect the mammalian fertilization pathway, as well as to identify and characterize zona pellucida components involved in the pathway. Application of these and other experimental approaches to zona pellucida synthesis, structure, and function continues to reveal the molecular bases of zona pellucida functions during mammalian development. It has also revealed some novel ways in which to address practical problems in mammalian conception and contraception.

This monograph, devoted exclusively to the zona pellucida, provides an up-to-date, comprehensive assessment of the field.

However, it certainly will not be the final word on the subject. Future research on the zona pellucida will undoubtedly reveal as yet unappreciated aspects of this unique organelle. I look forward to the next installment.

Paul M. Wassarman
Chairman
Department of Cell and Developmental Biology
Roche Institute of Molecular Biology

Preface

During growth of the ovarian follicle, the mammalian oocyte becomes surrounded by an acellular coat called the zona pellucida. Despite numerous morphological and biochemical studies, it has remained unresolved whether the zona pellucida originates from the oocyte, the follicle cells, or both. The past 10 years have seen the emergence of evidence supporting the idea that the zona pellucida originates from the oocyte itself. The zona pellucida mediates a number of important biological functions, including relatively species-specific binding of sperm to the ovulated eggs and postfertilization blocking of polyspermy. Sperm-zona pellucida interaction on a molecular basis is of inherent interest since it represents the first instance of cell-cell interaction during the development of a new individual.

In recent years the application of immunocytochemical and molecular biological techniques to the study of sperm-zona pellucida interaction has dramatically altered our concepts of gamete recognition. It is the aim of this book to present the reader with the most up-to-date information concerning the field of gamete interaction in reproductive biology. The molecular basis of the receptor-mediated process of gamete recognition is described in detail. In addition, there is extensive discussion of the site of synthesis of the extracellular egg coat, which has been a matter of dispute for decades. The resolution of this problem with the aid of the methods of molecular biology is given due consideration. There is also a description of the ultrastructural appearance during development of the zona pellucida in the various stages of follicle maturation. Furthermore, particular glycoproteins present in the egg coat are discussed and their possible role with regard to immunological contraception is examined closely. Finally, the zona pellucida is put forward as the possible site of origin of immunological infertility.

The subject matter of this book fills a gap in reproductive biology research and thereby provides a link between the separate scientific discussion of the egg and of the sperm.

Tübingen, December 1988 *Johannes Dietl*

Table of Contents

List of Contributors

Bleau, G.
Department of Obstetrics and Gynecology, Université de Montréal, Maisonneuve-Rosemont Research Center, 5415, boul. de l'Assomption, Montréal, Canada, H1T 2M4

Bleil, J.
Department of Cell and Developmental Biology, Roche Institute of Molecular Biology, Roche Research Center, Nutley, NJ 07110, USA

Chamberlin, M. E.
Laboratory of Cellular and Developmental Biology, Building 6, Room 81-08, NIDDK National Institutes of Health, Bethesda, MD 20892, USA

Chamow, S. M.
Laboratory of Cellular and Developmental Biology, Building 6, Room 81-08, NIDDK National Institutes of Health, Bethesda, MD 20892, USA

Dean, J.
Laboratory of Cellular and Developmental Biology, Building 6, Room 81-08, NIDDK National Institutes of Health, Bethesda, MD 20892, USA

Dietl, J.
Department of Obstetrics and Gynecology, Eberhard-Karls-Universität, Schleichstraße 4, D-7400 Tübingen, Federal Republic of Germany

Dunbar, B. S.
Department of Cell Biology, Baylor College of Medicine, One Baylor Plaza, Houston, TX 77030, USA

Fimiani, C.
Department of Cell and Developmental Biology, Roche Institute of Molecular Biology, Roche Research Center, Nutley, NJ 07110, USA

Florman, H.
Department of Cell and Developmental Biology, Roche Institute of Molecular Biology, Roche Research Center, Nutley, NJ 07110, USA

Greve, J.
Department of Cell and Developmental Biology, Roche Institute of Molecular Biology, Roche Research Center, Nutley, NJ 07110, USA

Gwatkin, R.B.L.
Reproductive and Developmental Biology, The Cleveland Clinic Foundation, 9500 Euclid Avenue, Cleveland, OH 44106, USA

Ida, K.
Department of Gynecology and Obstetrics, Faculty of Medicine, Kyoto University, Kyoto, Japan

Kinloch, R.
Department of Cell and Developmental Biology, Roche Institute of Molecular Biology, Roche Research Center, Nutley, NJ 07110, USA

Maresh, G.A.
Department of Cell Biology, Baylor College of Medicine, One Baylor Plaza, Houston, TX 77030, USA

Matsumoto, H.
Department of Gynecology and Obstetrics, Faculty of Medicine, Kyoto University, Kyoto, Japan

Moller, C.
Department of Cell and Developmental Biology, Roche Institute of Molecular Biology, Roche Research Center, Nutley, NJ 07110, USA

Mori, E.
Division of Immunology, Institute of Medical Science, Tokyo University, Tokyo, Japan

Mori, Takahide
Department of Gynecology and Obstetrics, Faculty of Medicine, Kyoto University, Kyoto, Japan

Mori, Tsuneatsu
Division of Immunology, Institute of Medical Science, Tokyo University, Tokyo, Japan

Mortillo, S.
Department of Cell and Developmental Biology, Roche Institute of Molecular Biology, Roche Research Center, Nutley, NJ 07110, USA

Nakano, M.
Department of Chemistry, Faculty of Science, Chiba University, Chiba 260, Japan

Noda, Y.
Department of Gynecology and Obstetrics, Faculty of Medicine, Kyoto University, Kyoto, Japan

Philpott, C. C.
Howard Hughes Medical Institute, Bethesda, MD 20814, USA

Ringuette, M. J.
Laboratory of Cellular and Developmental Biology, Building 6, Room 81-08, NIDDK National Institutes of Health, Bethesda, MD 20892, USA

Roller, R.
Department of Cell and Developmental Biology, Roche Institute of Molecular Biology, Roche Research Center, Nutley, NJ 07110, USA

Sacco, A. G.
Wayne State University School of Medicine, Department of Obstetrics and Gynecology, C. S. Mott Center for Human Growth and Development, 275 E. Hancock Avenue, Detroit, MI 48201, USA

Salzmann, G.
Department of Cell and Developmental Biology, Roche Institute of Molecular Biology, Roche Research Center, Nutley, NJ 07110, USA

St-Jacques, S.
Department of Obstetrics and Gynecology, Université de Montréal, Maisonneuve-Rosemont Research Center, 5415, boul. de l'Assomption, Montréal, Canada, H1T 2M4

Vazquez, M.
Department of Cell and Developmental Biology, Roche Institute of Molecular Biology, Roche Research Center, Nutley, NJ 07110, USA

Washenik, K.
Department of Cell Biology, Baylor College of Medicine,
One Baylor Plaza, Houston, TX 77030, USA

Wassarman, P.
Department of Cell and Developmental Biology, Roche Institute of Molecular Biology, Roche Research Center, Nutley, NJ 07110, USA

Yurewicz, E. C.
Wayne State University School of Medicine, Department of Obstetrics and Gynecology, C.S. Mott Center for Human Growth and Development, 275 E. Hancock Avenue, Detroit, MI 48201, USA

Molecular Genetics of the Mouse Zona Pellucida

M. E. Chamberlin, M. J. Ringuette, C. C. Philpott, S. M. Chamow, and J. Dean

Introduction

At fertilization male and female gametes fuse and initiate the development of a new organism. In mammals the nuclear material from both the sperm and the egg is ultimately required for embryogenesis to be successful (McGrath and Solter 1984; Surani et al. 1984). The two gametes, however, appear to make unequal contributions to the beginning embryo (Barton et al. 1984; Surani et al. 1986). Studies of parthenogenetically activated mouse eggs have demonstrated that the egg is capable of progressing to the egg cylinder stage (Kaufman et al. 1977), which suggests that the mature mouse oocyte contains sufficient information to direct the beginning of development. The nature of this maternal contribution has been studied extensively in some organisms, e.g., *Drosophila melanogaster, Xenopus laevis,* and the sea urchin (Konrad et al. 1985; Smith and Richter 1985; Brandhorst 1985; Raff and Showman 1985; Davidson 1986), but has been characterized to a lesser extent in mammals. What we do know about maternal inheritance in mammals has been gleaned primarily from studies of the laboratory mouse, which will be the focus of this chapter.

The genes that code for the mouse zona pellucida provide an ideal system in which to examine the role of genes expressed during mouse oogenesis. These genes code for three sulfated glycoproteins, ZP1, ZP2, and ZP3, which form a 7-μm thick extracellular matrix that surrounds the growing oocyte, the ovulated egg, and the dividing embryo (see Fig. 1). The expression of these genes is oocyte-specific and developmentally regulated. Synthesis of the protein products of the zona genes has been extensively investigated, and these protein products have well-defined biological functions that are crucial for early mammalian development. We will begin with an overview of mouse oogenesis with its attendant accumulation of macromolecules and then focus on the zona pellucida genes as specific examples of the expression of the maternal genome. The chapter will continue with a discussion of the structure and function of the protein products of the zona genes and finish with an examination of how some of their biological functions can be perturbed with anti-zona monoclonal antibodies.

Mouse Oogenesis

There is little direct evidence on the origin of the mouse germ line. Recent data using retroviral cell lineage markers suggest that the somatic lineages are derived from no more than eight founder cells in the inner cell mass, that the germ line is

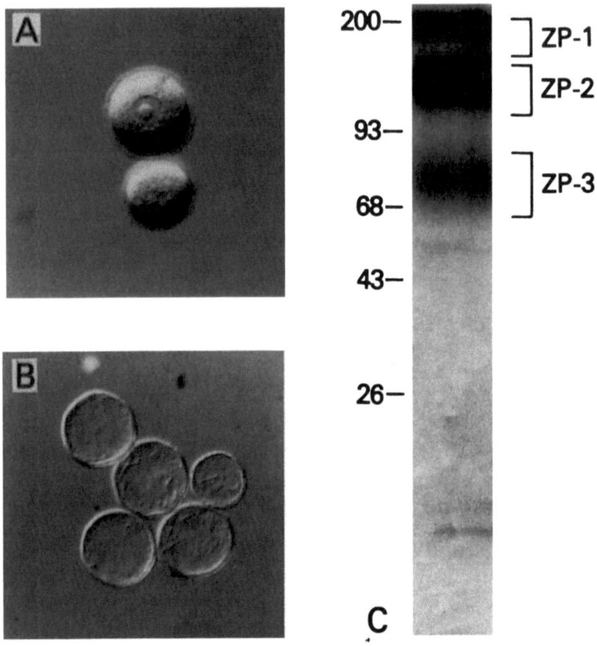

Fig. 1 A–C. Structure of the mouse zona pellucida. **A** Photomicrograph of growing oocytes isolated from juvenile mouse ovaries and grown in follicle culture. **B** Isolated intact zonae pellucidae. **C** ^{125}I-Labeled zona proteins separated by SDS-PAGE. (From Shimizu et al. 1983; East et al. 1984b)

derived from three cells, and that these are set aside prior to somatic tissue allocation (Soriano and Jaenisch 1986). Using morphological and histochemical criteria, primordial germ cells can first be detected 8 days post coitum in the allantoic region arising from the primitive streak (Chiquoine 1954; Mintz and Russell 1957; Ozdzenski 1967; Tam and Snow 1981). These early germ cells have locomotory activity and characteristics of invasive cells (Donovan et al. 1986). Over the next 4 days the primordial germ cells migrate to the gonadal ridges where they lose their motility and differentiate into male or female gonads (Blandau et al. 1963; Wylie et al. 1985; Donovan et al. 1986).

Initially no more than 15–75 primordial germ cells can be identified in the allantois, but they divide mitotically as they migrate, and more than 2500–6000 cells are detected in the developing gonad 12 days post coitum (Chiquoine 1954; Mintz and Russell 1957). Rapid mitosis continues, but oogonia gradually the preleptotene and leptotene stages of meiosis, and by day 16 (when there are several million germ cells) almost all oocytes are in the pachytene stage of the first meiotic division (Wassarman and Josefowicz 1978). The majority of these cells degenerate and the rest progress into the dictyate or resting phase of the first meiotic prophase. Thus, at birth, after 21 days of gestation, the mouse ovary contains roughly 10000–15000 oocytes (Jones and Krohn 1961) all of which are either in the dictyate or will become so within the first 5 days of life (Baker 1972). Oocytes continue to degenerate logarthmically during the life of the mouse,

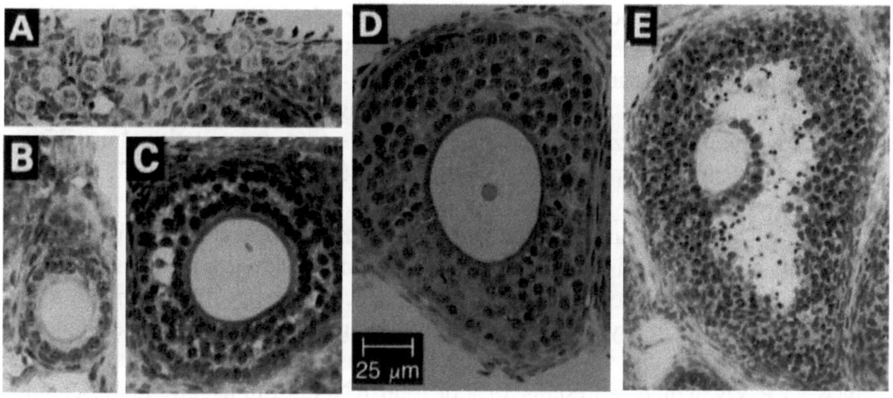

Fig. 2 A-E. Photomicrography of mouse ovaries stained with periodic acid-Schiff (PAS). **A** Primordial follicles containing resting oocytes and a single layer of spindle-shaped granulosa cells. **B** Small follicle with a growing oocyte surrounded by a thin, faintly PAS-staining zona pellucida and a layer of cuboidal granulosa cells. **C** Medium follicle with growing oocyte surrounded by PAS-staining zona and two to three layers of granulosa cells. **D** Large follicle with a fully grown oocyte and zona surrounded by multiple layers of granulosa cells. **E** Fully developed graafian follicle surrounding a grown oocyte. (From Philpott et al. 1987)

although a subset of this population is available (in most strains) to undergo the growth phase of oogenesis and subsequent ovulation (Jones and Krohn 1961; Pederson 1969). During growth the oocytes remain arrested in the first meiotic division, which is not completed until just prior to ovulation. The second meiotic division occurs at fertilization.

After birth successive cohorts of 10–15 oocytes enter into a 2-week growth phase during which time their diameters increase from 13 μm to 70 μm (Bramwell 1928). This increase occurs in conjunction with the proliferation of surrounding granulosa cells, creating follicles up to 125 μm in diameter (Fig. 2). During subsequent follicular growth to 600 μm the diameter of the oocyte increases little if at all (Bramwell 1928). The granulosa cells have specialized contacts with the oocyte and assist in protein and nucleic acid synthesis via metabolite transfer to the oocyte (Anderson and Albertini 1976; Gilula et al. 1978; Heller et al. 1981; Brower and Schultz 1982). During the growth phase of oogenesis the oocytes manufacture the majority of the components that they bring to the event of fertilization. Proteins that are known to be abundantly synthesized in the growing oocyte include tubulin (Schultz et al. 1979), lactate dehydrogenase (Brinster 1965; Mangia et al. 1976), actin and intermediate filaments (Kaplan et al. 1982), ribosomal proteins (La Marca and Wassarman 1979), and the proteins of the zona pellucida (Bleil and Wassarman 1980a; Shimizu et al. 1983). There are some qualitative changes in the two-dimensional gel pattern of proteins synthesized during the growth phase of oogenesis, and these differences become more pronounced during meiotic maturation and the completion of the first meiotic division (Schultz et al. 1979; Kaplan et al. 1982; Richter and McGaughey 1981). RNA synthesis is low or absent during meiotic maturation, and the appearance of new proteins is thus presumably due to the activation of stored maternal RNA (Bachvarova 1985). The second meiotic division occurs at fertilization, after which a number of changes in protein patterns have been observed (Levinson et al. 1978; Van Blerkom 1981; Cullen et al. 1980; Cascio and Wassarman 1982; Pratt et al. 1983). However, it appears that many, if not all, of the observed changes in protein patterns between oocytes and the one cell zygote (see above) are due to posttranslational modifications of preexisting polypeptides (Van Blerkom 1981; Pratt et al. 1983; Howlett 1986).

RNAs of all classes are also actively synthesized and accumulate during the growth phase of oogenesis (for reviews, see Bachvarova 1985; Davidson 1986). Of particular interest are the mRNAs, some of which may be utilized in the early development of the embryo before the embryonic genome is able to direct its own RNA synthesis. By the end of its growth phase, the oocyte contains 90 pg poly(A+) RNA, which represents 19% of total RNA, an unusually high percentage compared to the 1%–5% poly(A+) RNA found in somatic cells (Bachvarova et al. 1985). Approximately 75% of this polyadenylated RNA is stored as stable maternal message, while the rest turns over with the relatively slow half-life of 6 days (DeLeon et al. 1983). However, during meiotic maturation and ovulation approximately half the poly(A+) RNA disappears. Some of this loss is due to deadenylation (Bachvarova et al. 1985), and some may be due to degradation of mRNAs. Change in the stability of mRNAs appears to be due to changes in the cytoplasm, in that similar half-lives have been observed with heterologous mRNAs microinjected into oocytes and fertilized eggs (Ebert and Brinster 1983; Ebert et al. 1984). By the two-cell embryo stage (40 hours after fertilization) less than 10% of the original maternal mRNA is still present (Clegg and Piko 1983), and the embryonic genome is clearly active (Flach et al. 1982; Bensaude et al. 1983; Giebelhaus et al. 1983; Bolton et al. 1984; Howlett 1986).

Genomic Organization of ZP3

To investigate how maternally encoded information is utilized in early development, we have chosen to study the molecular details of the genes that code for the mouse zona pellucida proteins. These proteins are synthesized only by the oocyte genome and persist with known biological functions at fertilization and in early development. Using a monoclonal antibody specific to ZP3 as a probe (East et al. 1985), a cDNA coding for this protein was isolated from an ovarian lambda gt11 expression library. The identity of the cDNA clone was confirmed by comparing a translation of its nucleic acid sequence with the amino acid sequence of an internal peptide derived from the ZP3 protein (Ringuette et al. 1986; see Fig. 3). This cDNA has been used to answer some questions about the genomic organization of this specific zona protein gene.

The gene which encodes ZP3 is present in a single copy in the mouse genome (Ringuette et al. 1986; Chamberlin and Dean, unpublished observations). Hybridization of the ZP3 cDNA to DNA from 15 different Chinese hamster ovary cell-mouse somatic cell hybrid tissue culture lines demonstrates that ZP3 is located on chromosome 6 of the mouse (Philpott et al., unpublished observations). A restriction fragment length polymorphism (RFLP) was detected at the ZP3 locus

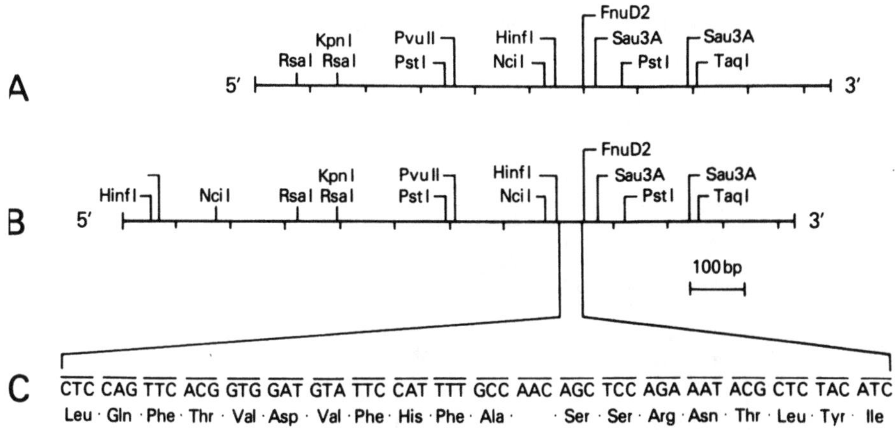

Fig. 3 A–C. Restriction maps and partial sequence of ZP3 cDNA clones. **A** Restriction map of pZP-3.1 cDNA insert. **B** Restriction map of pZP-3.2 cDNA insert aligned with pZP-3.1. **C** Comparison of the cDNA sequence of a portion of pZP-3.1 with the amino acid sequence of a peptide isolated by antibody affinity chromatography from a V-8 protease digest of the ZP3 protein. An asparagine residue which would fall in the *N*-glycosylation sequence, Asn-X-Ser, does not appear in the amino acid sequence because derivatized Asn residues are not detected by the procedures used. (From Ringuette et al. 1986)

using DNAs from different recombinant inbred strains of mice after digestion with the restriction enzyme Pst I. The different strains contain one of at least two alleles at this locus (Philpott et al., unpublished observations). However, there is not a sufficiently close genetic linkage of the ZP3 RFLP to any of the five known genetic markers on chromosome 6 (MTV, ENV-27, LY-2, LVP, and GGC) for more exact localization of the ZP3 locus by this approach.

Interestingly, the gene that encodes the species-specific sperm receptor for the mouse appears to have homologues in other mammalian species (Ringuette et al. 1986). Using a relatively stringent criterion for hybridization, one that should detect homologies of greater than 78%, cross-hybridization can readily be seen with the DNA from other mammals, such as rat, dog, cow, and human, and less well with DNA from rabbit and pig. However, the cross-hybridization signal of these latter two species increases qualitatively and quantitatively at reduced stringency (see Fig. 4). The homology of the sequence of the ZP3 locus at the genetic level may reflect the common function of the sperm receptor among mammalian species. It has been shown that the O-linked oligosaccharide moieties of ZP3 are required for sperm receptor activity (Florman et al. 1984; Florman and Wassarman 1985), and it may be these modifications of the core protein that account for the species-specific interaction of sperm and egg at fertilization.

Little is known about mechanisms that modulate gene expression during mammalian oogenesis. As will be described below, transcription of the ZP3 gene occurs only during the growth of oocytes arrested in the prophase of the first meiotic division. Thus, the turning on and the turning off of the ZP3 gene occurs in the absence of DNA replication and cell division. The gene apparently undergoes no rearrangement or amplification in order to be expressed (Ringuette et al. 1986).

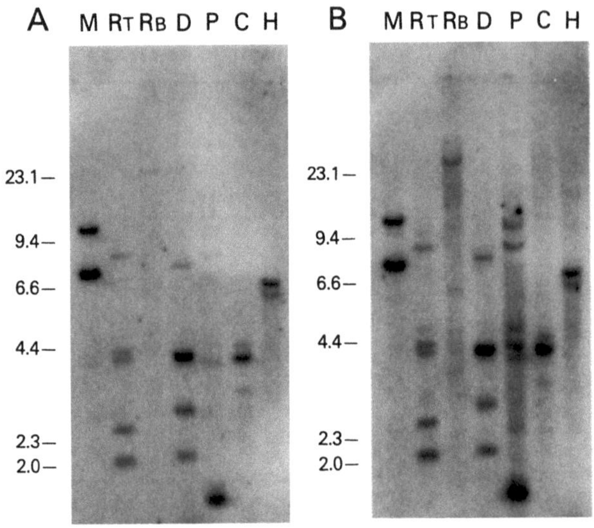

Fig. 4 A, B. Cross-hybridization of ZP3 among mammals. **A** Southern blot of genomic DNA from mouse *(M)*, rat *(Rt)*, rabbit *(Rb)*, dog *(D)*, pig *(P)*, cow *(C)*, and human *(H)*, probed with ^{32}P-labeled mouse ZP3 cDNA and washed under conditions in which homologies of >78% should be detected. **B** Same as **A** but washed under reduced stringency in which homologies of >54% should be detected. (From Ringuette et al. 1986)

It has been observed that the DNA of the ZP3 locus is hypomethylated in ovarian tissue, compared to DNA isolated from somatic tissues such as brain or liver (Ringuette et al. 1986). DNA in the mouse has been shown to be extensively methylated in most tissues (Chapman et al. 1984; Jahner and Jaenisch 1984) although hypomethylation is observed in oocytes at single-copy gene loci as well as in DNA containing repetitive elements (Sanford et al. 1984, 1985). Thus, the functional significance of hypomethylation at the ZP3 locus, and how it may relate to the tissue-specific expression of ZP3 remains to be determined.

Developmental Expression of ZP3

Northern blot analysis demonstrated that ZP3 is transcribed only in ovarian tissue, where it is expressed as a 1.5- to 1.6-kb poly(A+) transcript (Ringuette et al. 1986; see Fig. 5). In addition to a poly A tail, the ZP3 mRNA contains 1308 nucleotides and has a single 1270-nucleotide open reading frame which is flanked by relatively short 5′ and 3′ untranslated regions. The deduced amino acid sequence codes for a 424 amino acid polypeptide chain of 46617 daltons and contains a potential signal peptide which would result in a secreted polypeptide chain of 43949 daltons (Ringuette et al., unpublished observations). The protein also has six potential N-linked glycosylation sites, three or four of which appear to be glycosylated in the mature ZP3 protein (Salzmann et al. 1983).

In situ hybridization of an anti-sense ZP3 RNA to methacrylate embedded sections of ovaries detects ZP3 transcripts only in growing oocytes. No ZP3 tran-

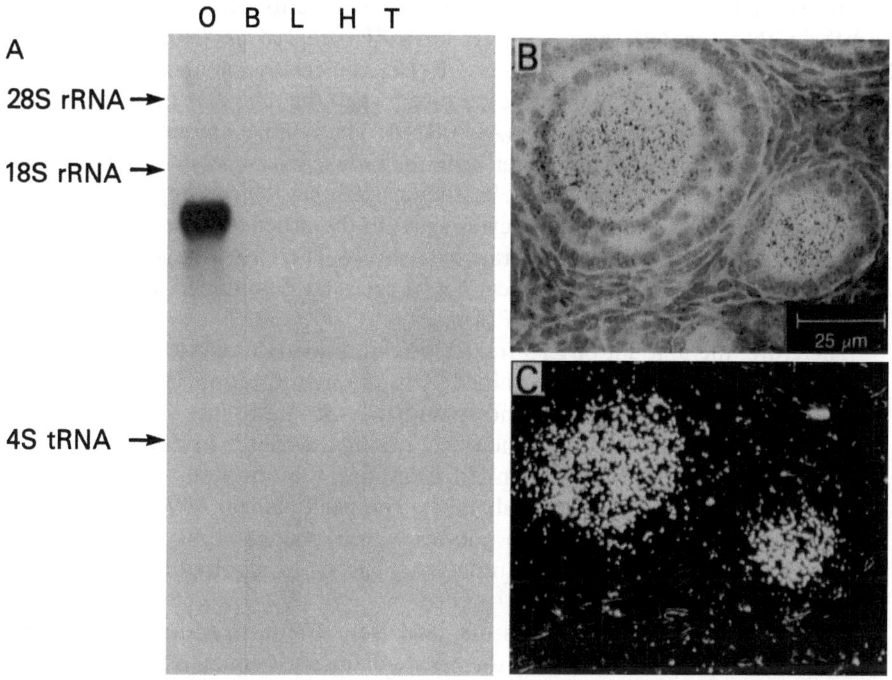

Fig. 5 A–C. Tissue-specific expression of ZP3. **A** Northern blot analysis of RNA from different mouse tissues probed with a ^{32}P-labeled ZP3 insert. *Lane O,* 2-μg ovarian poly(A+) RNA; *lane L,* 10-μg total liver RNA; *lane H,* 10-μg total heart RNA; *lane B,* 10-μg total brain RNA; *lane T,* 10-μg total testes RNA. **B** ^{35}S-labeled anti-sense RNA transcripts derived from the ZP3 cDNA were hybridized to methacrylate embedded ovarian sections and stained with Giemsa. **C** Same as **A** but viewed with darkfield prior to staining. (From Ringuette et al. 1986; Philpott et al. 1987)

scripts were detected in the surrounding granulosa cells either by in situ hybridization of ovarian tissue or by Northern blot analysis of isolated granulosa cells (Philpott et al. 1987; see Fig. 5). Thus, ZP3 expression appears to be oocyte-specific and limited to the growth phase of oogenesis.

Morphologically, no zona pellucida can be detected around oocytes prior to entry into the growth phase of oogenesis, and ZP3 RNA transcripts are not detected in the poly(A+) RNA from the prenatal ovaries which contain only resting oocytes (10–15 μm diameter; Philpott et al. 1987). During the first 3 weeks of life ovaries of juvenile mice contain (in addition to resting oocytes) relatively homogeneous cohorts of growing oocytes, such that ovaries from 1-, 3-, 9-, and 15-day-old mice are primarily populated with growing oocytes whose diameters are 15 μm, 25 μm, 35 μm, and 50 μm, respectively (Pedersen 1969; Peters 1969; Mangia and Epstein 1975). Poly(A+) RNAs isolated from ovaries of mice at different ages show that the abundance of ZP3 transcripts as a percentage of total ovarian poly(A+) RNA increases with age until day 9, at which time its relative abundance begins to decrease (Philpott et al. 1987). The number of transcripts were quantified relative to a known number of sense-strand ZP3 transcripts made in vitro from a plasmid containing a promoter specific for RNA polymerase SP6 and the cDNA for ZP3. At 1 day after birth ZP3 transcripts are detected at the level of 45 pg/μg of ovarian poly(A+) RNA. The relative abundance increases until maximal expression at 9 days after birth, when the amount of ZP3 transcript is equal to 300 pg/μg ovarian poly(A+) RNA. The relative amount of ZP3 message then decreases. At 21 days after birth the mouse ovary poly(A+) RNA contains 60 pg ZP3 mRNA/ug. It must be emphasized that the cell population of the ovary is mixed, and that as the oocytes grow in the absence of cell division, the surrounding granulosa cells proliferate extensively. Therefore, the absolute abundance of ZP3 transcript per oocyte cannot be precisely determined from Northern blot analysis of ovarian poly(A+) RNAs.

Therefore, the amount of ZP3 transcribed in individual oocytes as a function of increasing oocyte size was determined by in situ hybridization. Ovarian sections were embedded in methacrylate and hybridized with anti-sense ZP3 RNA. The grain density was analyzed and standardized using dot-blot hybridization to RNA from medium-sized oocytes (35 μm) to determine the absolute amount of ZP3 transcripts per oocyte (Philpott et al. 1987). The peak content of ZP3 messengers (220 fg per oocyte) was detected in oocytes with diameters of 50 μm. This represents 0.1%-0.2% of the total ovarian poly(A+) RNA as reported in the literature (DeLeon et al. 1983).

Thus, as an oocyte increases in diameter from 10 μm to 65 μm, ZP3 is transcribed in increasing amounts until the oocyte diameter is equal to 50 μm; thereafter the amount of ZP3 transcript decreases substantially. In a fully grown oocyte (65 μm using this embedding system) ZP3 transcripts are present at 65% of maximal expression. Interestingly, a small amount of ZP3 transcript (15%-20% of maximal expression) can be detected in ovulated eggs (Philpott et al. 1987) although no zona synthesis is detectable at this time (Bleil and Wassarman 1980b; Shimizu et al. 1983). It remains to be determined whether the ZP3 transcripts present at this time are partially degraded or fully functional.

Biosynthesis of the Zona Pellucida Matrix

There has been considerable controversy whether the zona proteins are synthesized uniquely in oocytes or in both oocytes and granulosa cells. Some morphological investigations suggested that, in addition to oocytes, the surrounding granulosa cells produce zona proteins (Chiquoine 1960; Hadek 1965), and this contention is supported by immunohistochemical work using a polyclonal anti-zona antiserum (Wolgemuth et al. 1984). However, other morphological studies suggest that the zona is produced only in oocytes (Kang 1974; Haddad and Nagai 1977; Bousquet et al. 1981; Flechon et al. 1984). In vitro labeling experiments indicate that denuded oocytes (but not follicle cells) are capable of synthesizing all three zona proteins (Bleil and Wassarman 1980b). Our more recent studies, using anti-sense RNA probes, also detected ZP3 transcripts only in growing oocytes and not in the surrounding granulosa cells (Philpott et al. 1987). Thus, it appears most likely that the zona proteins are expressed in an oocyte-specific manner.

The rate of zona protein biosynthesis has been measured in a follicle culture system using an isotope dilution method (Shimizu et al. 1983). As the oocytes grow in diameter from roughly 50 μm to 65 μm, the fraction of total protein synthesis devoted to producing zona proteins drops from 7% to 2% (Shimizu et al. 1983); this corresponds qualitatively to the decrease observed in the abundance of ZP3 transcripts during the later stages of oocyte growth (Philpott et al. 1987). Fully grown oocytes do not appear to make zona proteins (Bleil and Wassarman 1980b), and oocytes that have been superovulated, although still synthesizing protein, no longer synthesize zona proteins (Shimizu et al. 1983). Thus, the rise and fall of zona protein synthesis is closely correlated with the abundance of zona gene transcripts. These observations suggest, but do not rigorously demonstrate, that zona production is regulated at a transcriptional level.

Each of the three zona proteins is a sulfated glycoprotein, as are other extracellular matrix proteins (Shimizu et al. 1983). The zona proteins contain both N-linked and O-linked oligosaccharide side chains. The largest of the zona proteins is ZP1 with an apparent molecular weight of 200 kd (Bleil and Wassarman 1980a; Shimizu et al. 1983). ZP2 appears to consist of a core protein of 81 kd and six N-linked oligosaccharide side chains (Greve et al. 1982) which together with uncharacterized O-linked sugars form a mature secreted protein of 120 kd. ZP3 (secreted as a 85-kd protein) has been reported to be synthesized as a polypeptide of 44 kd to which either three or four N-linked oligosaccharides are attached (Salzmann et al. 1983). The nucleic acid sequence of the cDNA coding for ZP3 indicates the presence of six potential Asn-X-Ser/Thr sites for glycosylation of N-linked sugar side chains (Ringuette et al., unpublished observations).

The zona pellucida surrounding a fully grown mouse oocyte is 7 μm thick and contains 3-4 ng of protein, made up of ZP1, ZP2 and ZP3 (Bleil and Wassarman 1980a; Shimizu et al. 1983). The three proteins appear to be organized in a matrix of branched filaments (Greve and Wassarman 1985). Certain chemical and enzymatic treatments, specifically dithiothreitol and chymotrypsin, reduce the branched matrix to individual filaments that appear to be composed of both ZP2 and ZP3, suggesting that ZP1 is responsible for the branching (Greve and Wassarman 1985). Thus, the zona pellucida of a fully grown oocyte appears to be an ordered structure of filaments composed of ZP2 and ZP3, either together or separate, and held together by ZP1. The entire structure undergoes a change upon fertilization (the zona reaction) which prevents polyspermy. This may be due to a biochemical alteration in one or all of the zona proteins that changes the three dimensional structure of the matrix.

Pulse-chase studies indicate that the zona proteins have an unusually long half-life in vitro (Shimizu et al. 1983). This is consistent with the observation that the mouse zona persists until implantation, which occurs 4.5 days after fertilization. This persistence is important for the biological functions of the zona, all of which occur after zona protein synthesis has finished (see below). At the time of normal implantation the mouse blastocyst hatches out of the zona pellucida in vitro using a serine proteinase located in the mural trophectoderm (Perona and Wassarman 1986).

Functions of the Zona Pellucida

The zona pellucida has a number of biological functions that are vital to early mammalian development. The zona mediates the species-specific sperm-egg interaction at fertilization (Yanagimachi 1981; Wassarman 1987), and following fertilization, presumably by a modification of its structure, it serves as a major block to polyspermy (Wolf 1981; Wassarman 1987). The zona also serves to protect the very young embryo as it passes down the oviduct and persists until the embryo implants in the uterine wall.

At fertilization acrosome-intact sperm bind to mouse zona; this interaction can be inhibited by solubilized zonae isolated from ovulated eggs but not from one-cell zygotes (Gwatkin and Williams 1976; Bleil and Wassarman 1980c). This inhibition has been ascribed to a class of 3.9-kd O-linked oligosaccharide side chains of ZP3, which suggests that these sugar side chains act as the initial sperm receptor (Florman and Wassarman 1985). The process of homologous sperm binding to the zona occurs in two steps: loose attachment followed by strong binding (Saling et al. 1979; Florman and Storey 1982). It appears that subsequent to these two events, the presence of the polypeptide backbone of ZP3 is necessary for the triggering of the acrosome reaction (Florman et al. 1984) - a vesiculation of the sperm's outer acrosomal membrane which is a prerequisite for sperm penetration of the zona pellucida. Acrosome-reacted sperm then appear to bind to ZP2, which has recently been postulated to act as a secondary receptor (Bleil and Wassarman 1986). Acrosin, a serine proteinase located in the sperm acrosome, is thought to play an important role in the penetration of the zona by the sperm.

After the fusion of the sperm and egg cell membranes the sperm is incorporated into the egg and cortical granule release occurs. ZP3 purified from two-cell embryos is no longer capable of inducing the acrosome reaction (Bleil and Wassarman 1983) or of interfering with sperm binding (Bleil and Wassarman 1980c); this suggests that fertilization causes a change in ZP3 that renders it incapable of these two functions. Although the biochemical basis of this change has yet to be elucidated, enzymes released from the cortical granules at fertilization may be at least partly responsible. Recent data suggests that activators of protein kinase C modify the zona pellucida so that it no longer induces the acrosome reaction but still has sperm binding ability (Endo et al. 1987). The molecular dissection of these two vital processes should permit further elucidation of the molecular basis of fertilization.

ZP2 has also been shown to undergo a significant modification upon fertilization (Bleil et al. 1981). Its apparent molecular weight decreases by 30 kd, a change which is only apparent after reduction of disulfide bonds. This suggests that ZP2 is proteolytically cleaved at fertilization, but the polypeptide fragments are still attached via intramolecular disulfide bonds (Bleil et al. 1981). Cleavage of ZP2 in this manner may allow for a change in the three-dimensional structure of the entire zona (the zona reaction) which would render it impermeable to additional sperm and account for the postfertilization block to polyspermy.

Perturbation of Zona Function with Antibodies

Administration of anti-zona antibodies to mice produces temporary infertility (Sacco 1979; Tsunoda et al. 1981; Aitken et al. 1981) without otherwise affecting early development (Tsunoda and Whittingham 1982). Similar results have been obtained in other mammals following active immunization with ovary homogenates or isolated zona pellucida (Gwatkin et al. 1977; Wood et al. 1981; Mahi-Brown et al. 1982; Gulyas et al. 1983). The more recent availability of well-defined monoclonal antibodies specific to unique epitopes on individual mouse zona proteins has permitted an extension of this early work (Dean et al. 1986a).

Rat anti-mouse ZP2 and ZP3 monoclonal antibodies have been characterized and have been passively administered to mice both before and after fertilization (East and Dean 1984; East et al. 1984a, 1985). After administration these antibodies localize uniquely to the ovaries (East et al. 1984b) where they bind only to growing oocytes (East and Dean 1984; East et al. 1985; see Fig. 6). The antibodies occupy roughly 1.3×10^8 sites per oocyte and have dissociation constants which range from 6×10^{-10} to 3×10^{-9} mol/l. The binding of antibodies to the zona persists after ovulation and the presence of these antibodies precludes fertilization (East et al. 1984a, 1985). Antibodies specific to ZP2 or to ZP3 are equally effective

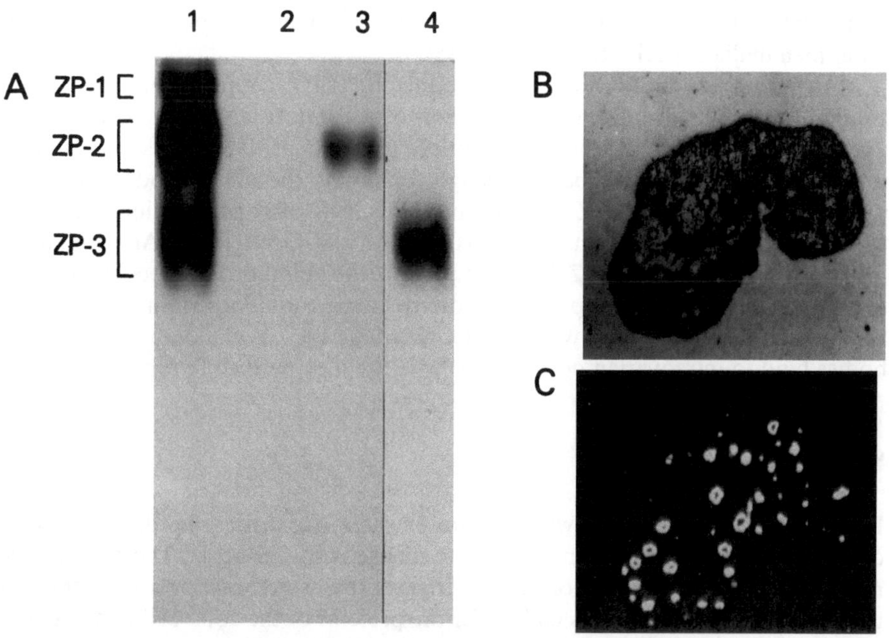

Fig. 6 A-C. Anti-zona pellucida monoclonal antibodies. **A** Immunoprecipitation of ^{125}I-labeled zona followed by SDS-PAGE and autoradiography *Lane 1,* zona pellucida; *lane 2,* immunoprecipitation with ascites fluid; *lane 3,* immunoprecipitation with an anti-ZP2 monoclonal antibody; *lane 4,* immunoprecipitation with an anti-ZP3 monoclonal antibody. **B** Methylene-blue stained frozen ovarian section obtained from a mouse passively immunized with anti-ZP2 monoclonal antibodies. **C** Indirect immunofluorescence of **B** with rhodamine-conjugated anti-rat antibodies. (From Dean et al. 1986a)

as contraceptive agents. In vitro studies demonstrated that anti-ZP2 or -ZP3 antibodies did not inhibit fertilization by preventing sperm from binding to ovulated eggs but rather by blocking sperm penetration through the zona pellucida (East et al. 1985).

The in vivo contraceptive effect was long-lasting but eventually reversible; this observation can be explained in the following manner (East et al. 1984a). After administration the anti-zona monoclonal antibodies bind to the zona surrounding oocytes at different stages of growth. Resting oocytes have not yet begun to synthesize zona proteins, and therefore this pool of immature oocytes is not coated (see above). It appears that as long as cohorts of growing oocytes (coated with antibody) are ovulated, the animal remains infertile. However, as the circulating titer of antibodies declines, new oocytes synthesize zona in the absence of anti-zona antibodies. When ovulated, these oocytes can be fertilized and the contraceptive effect is reversed.

Anti-zona monoclonal antibodies administered after fertilization had no discernible effect on early development. Two-cell embryos grew normally to blastocysts in vitro and eventually hatched out of the antibody-coated zona (East et al. 1985). Embryos exposed to antibodies in vivo at the two-cell stage or later developed normally to term (East et al. 1984a). The ovarian histology of the antibody-treated animals appeared normal, and there was no destruction of intraovarian oocytes after a one-treatment trial (East et al. 1984a). All of the monoclonal antibodies reacted with both mouse and rat zona, but they did not with zona from other mammalian species (East et al. 1985).

The recent isolation of a cDNA coding for ZP3, coupled with the above studies, has now permitted the exploration of a contraceptive strategy based on active immunization with zona polypeptides (Dean et al. 1986b). *Escherichia coli* infected with lambda gt11 bacteriophage containing the cDNA coding for ZP3 produce a β-galactosidase/ZP3 fusion protein which, after purification, was used to immunize a series of female mice (Chamow and Dean 1987). After multiple injections with homologous ZP3 fusion protein, the animals developed antibodies that cross-reacted with ZP3 purified under reducing conditions from mouse zona. Studies are currently under way to determine whether circulating anti-zona antibodies bind to intraovarian oocytes, and whether they perturb fertilization.

Summary

From these studies the following picture emerges concerning the developmental expression and biological function of the mouse zona pellucida. The commencement of growth by a cohort of oocytes triggers (by as yet unknown mechanisms) the transcription of the genes coding for the proteins of the zona pellucida. As the oocytes grow, the zona mRNAs accumulate until ZP3 transcripts represent 0.1%-0.2% of the total mRNA (where the oocyte is approximately 50% of its full growth volume). At this point it is estimated that almost 10% of the considerable protein synthesis in the oocyte is devoted to zona production. The subsequent decrease in the abundance of zona transcripts closely follows the observed decrease in protein synthesis and secretion. No protein synthesis is detected after ovulation although low levels of ZP3 transcripts persist. The synthesized zona proteins are secreted to form a stable extracellular matrix which effectively maintains the integrity of the species by acting as the species-specific sperm receptor and by preventing postfertilization polyspermy.

Thus, the genes coding for the proteins of the mouse zona pellucida represent an interesting system for studies on the control of gene expression. The genes for these proteins are among the best examples of tissue-specific expression known. They are expressed only in germ line cells and then only in the female germ line cells. Their expression is limited to a very precise period of time during the 2-week growth phase of oogenesis when oocytes are arrested in the prophase of the first meiotic division. No cell division occurs during the entire time of developmental regulation of the zona genes. The highly specific nature of this expression suggests that a set of equally specific regulators must exist for the control of these genes. The molecular nature of these regulatory elements is currently under investigation.

Acknowledgements. We thank Drs. F.D. Costantini, J.L. Hedrick, and R.T. Simpson for critical review of the manuscript. One of the authors, C.C.P., is a Howard Hughes Medical Institute-National Institutes of Health Research Scholar.

References

Aitken RJ, Rudak EA, Richardson DW, Dor J, Djanhanbahkck O, Templeton AA (1981) The influence of antizona and antisperm antibodies on sperm-egg interactions. J Reprod Fertil 62: 597-606

Anderson E, Albertini DF (1976) Gap junctions between the oocyte and companion follicle cells in the mammalian ovary. J Cell Biol 71: 680-686

Bachvarova R (1985) Gene expression during oogenesis and oocyte development in mammals. In: Browder L (ed) Developmental biology: A comprehensive synthesis, vol 1. Plenum, New York, pp 453-524

Bachvarova R, DeLeon V, Johnson A, Kaplan G, Paynton BV (1985) Changes in total RNA, polyadenylated RNA and actin mRNA during meiotic maturation of mouse oocytes. Dev Biol 108: 325-331

Baker TG (1972) Oogenesis and ovulation. In: Austin CR, Short RV (eds) Reproduction in mammals: Germ cells and fertilization, vol 1. Excerpta Medica, Amsterdam, pp 398-437

Barton SC, Surani MAH, Norris ML (1984) Role of paternal and maternal genomes in mouse development. Nature 311: 374-376

Bensaude O, Babinet C, Morange M, Jacob F (1983) Heat shock proteins, first major products of zygotic gene activity in mouse embryo. Nature 305: 331-333

Blandau RJ, White BJ, Rumery RE (1963) Observations on the movements of the living primordial germ cells in the mouse. Fertil Steril 14: 482-489

Bleil JD, Wassarman PM (1980a) Structure and function of the zona pellucida: identification and characterization of the proteins of the mouse oocyte's zona pellucida. Dev Biol 76: 185-202

Bleil JD, Wassarman PM (1980b) Synthesis of zona pellucida proteins by denuded and follicle-enclosed mouse oocytes during culture in vitro. Proc Natl Acad Sci USA 77: 1029-1033

Bleil JD, Wassarman PM (1980c) Mammalian sperm-egg interaction: identification of a glycoprotein in mouse egg zonae pellucidae possessing receptor activity for sperm. Cell 20: 873-882

Bleil JD, Wassarman PM (1983) Sperm egg interactions in the mouse: sequence of events and induction of the acrosome reaction by a zona pellucida glycoprotein. Dev Biol 95: 317-324

Bleil JD, Wassarman PM (1986) Autoradiographic visualization of the mouse egg's sperm receptor bound to sperm. J Cell Biol 102: 1363-1371

Bleil JD, Beall CF, Wassarman PM (1981) Mammalian sperm egg interaction: fertilization of mouse eggs triggers modification of the major zona glycoprotein ZP2. Dev Biol 86: 189-197

Bolton VN, Oades PJ, Johnson MH (1984) The relationship between cleavages, DNA replication and gene expression in the mouse 2-cell embryo. J Embryol Exp Morphol 79: 139-163
Bousquet D, Leveille MC, Roberts KD, Chapdelaine A, Bleau G (1981) The cellular origin of the zona pellucida antigen in the human and hamster. J Exp Zool 215: 215-218
Bramwell FWR (1928) The development and morphology of the gonads of the mouse. Part III. Proc R Soc Lond [Biol] 103: 258-272
Brandhorst BP (1985) Informational content of the echinoderm egg. In: Browder L (ed) Developmental biology: A comprehensive synthesis, vol 1. Plenum, New York, pp 525-576
Brinster R (1965) Lactate dehydrogenase activity in the preimplanted mouse embryo. Biochim Biophys Acta 110: 439-441
Brower PT, Schultz RM (1982) Biochemical studies of mammalian oogenesis: possible existence of a ribosomal and poly(A)-containing RNA-protein supramolecular complex in mouse oocytes. J Exp Zool 220: 251-260
Cascio SM, Wassarman PM (1982) Development in the mammal: post transcriptional control of a class of proteins synthesized by mouse oocytes and early embryos. Dev Biol 89: 397-408
Chamow SM, Dean J (1987) Anti-zona pellucida antibodies in mice immunized with recombinant ZP3/β-galactosidase fusion protein. Fed Proc 46: 2015
Chapman V, Forrester L, Sanford J, Hastek H, Rossant J (1984) Cell lineage-specific undermethylation of mouse repetitive DNA. Nature 307: 284-286
Chiquoine AD (1954) The identification, origin and migration of the primordial germ cells in the mouse embryo. Anat Rec 118: 135-146
Chiquoine AD (1960) The development of the zona pellucida of the mammalian ovum. Am J Anat 106: 149-169
Clegg KB, Piko L (1983) Poly(A) length, cytoplasmic adenylation and synthesis of poly(A)+ RNA in early mouse embryos. Dev Biol 95: 331-341
Cullen B, Emigholz K, Monahan J (1980) The transient appearance of specific proteins in 1-cell mouse embryos. Dev Biol 76: 215-221
Davidson EH (1986) Gene activity in early development, 3rd edn. Academic, Orlando
Dean J, East IJ, Shimizu S (1986a) Biosynthesis of the mouse zona pellucida and the effect of anti-zona monoclonal antibodies on fertilization and early development. Theriogenology 25: 107-116
Dean J, Ringuette M, Sobieski DA, East IJ, Chamow S (1986b) Molecular genetics of the mouse zona pellucida: implications for fertilization and early development. In: Talwar GP (ed) Immunological approaches to contraception and promotion of fertility. Plenum, New York, pp 241-249
DeLeon V, Johnson A, Bachvarova R (1983) Half lives and relative amounts of stored and polysomal ribosomes and poly(A)+ RNA in mouse oocytes. Dev Biol 98: 400-408
Donovan PJ, Stott D, Cairns LA, Heasman J, Wylie CC (1986) Migratory and postmigratory mouse primordial germ cells behave differently in culture. Cell 44: 831-838
East IJ, Dean J (1984) Monoclonal antibodies as probes of the distribution of ZP2, the major sulfated glycoprotein of the murine zona pellucida. J Cell Biol 98: 795-800
East IJ, Mattison DR, Dean J (1984a) Monoclonal antibodies to the major protein of the murine zona pellucida: effects in fertilization and early development. Dev Biol 104: 49-56
East IJ, Keenan AM, Larson SM, Dean J (1984b) Scintigraphy of normal mouse ovaries with monoclonal antibodies to ZP2, the major zona pellucida protein. Science 225: 938-940
East IJ, Gulyas BJ, Dean J (1985) Monoclonal antibodies to the murine zona pellucida protein with sperm receptor activity: effects on fertilization and early development. Dev Biol 109: 268-273
Ebert KM, Brinster RL (1983) Rabbit α-globin messenger RNA translation by the mouse ovum. J Embryol Exp Morphol 74: 159-168
Ebert KM, Paynton BV, McKnight GS, Brinster RL (1984) Translation and stability of ovalbumin messenger RNA injected into growing oocytes and fertilized ova of mice. J Embryol Exp Morphol 84: 91-103
Endo Y, Schultz RM, Kopf GS (1987) Effects of phorbol esters and diacylglycerol on mouse eggs: inhibition of fertilization and modification of the zona pellucida. Dev Biol 119: 199-209
Flach G, Johnson MH, Braude PR, Taylor R, Bolton VN (1982) The transition from maternal to embryonic control in the 2-cell mouse embryo. EMBO J 1: 681-686

Flechon JE, Pavlok A, Kopecny V (1984) Dynamics of zona pellucida formation by mouse oocyte. An autoradiographic study. Biol Cell 51: 403-406

Florman HM, Storey BT (1982) Mouse gamete interactions: the zona pellucida is the site of the acrosome reaction leading to fertilization in vitro. Dev Biol 91: 121-130

Florman HM, Wassarman PM (1985) O-linked oligosaccharides of mouse egg ZP3 account for its sperm receptor activity. Cell 41: 313-324

Florman HM, Bechtol KB, Wassarman PM (1984) Enzymatic dissection of the functions of the mouse egg's receptor for sperm. Dev Biol 106: 243-255

Giebelhaus DH, Heikkila JJ, Schultz GA (1983) Changes in the quantity of histone and actin messenger RNA during the development of preimplantation mouse embryos. Dev Biol 98: 148-154

Gilula NB, Epstein ML, Beers WH (1978) Cell-to-cell communication and ovulation. A study of the cumulus-oocyte complex. J Cell Biol 78: 58-75

Greve JM, Wassarman PM (1985) Mouse egg extracellular coat is a matrix of interconnected filaments possessing a structural repeat. J Mol Biol 181: 253-264

Greve JM, Salzmann GS, Roller RJ, Wassarman PM (1982) Biosynthesis of the major zona pellucida glycoprotein secreted by oocytes during mammalian oogenesis. Cell 31: 749-759

Gulyas BJ, Gwatkin RBL, Yuan LC (1983) Active immunization of cynomolgus monkeys *(Macaca fascicularis)* with porcine zonae pellucidae. Gamete Res 4: 299-307

Gwatkin RBL, Williams DT (1976) Receptor activity of the hamster and mouse solubilized zona pellucida before and after the zona reaction. J Reprod Fertil 49: 55-59

Gwatkin RBL, Williams DT, Carlo DJ (1977) Immunization of mice with heat-solubilized hamster zonae: production of anti-zona antibody and inhibition of fertility. Fertil Steril 28: 871-877

Haddad A, Nagai EMT (1977) Radioautographic study of glycoprotein biosynthesis and renewal in the ovarian follicles of mice and the origin of the zona pellucida. Cell Tissue Res 177: 347-369

Hadek R (1965) The structure of the mammalian egg. Int Rev Cytol 18: 29-68

Heller DT, Cahill DM, Schultz RM (1981) Biochemical studies of mammalian oogenesis: metabolic cooperativity between granulosa cells and growing mouse oocyte. Dev Biol 84: 455-464

Howlett SK (1986) A set of proteins showing cell cycle dependent modification in the early mouse embryo. Cell 45: 387-396

Jahner D, Jaenisch R (1984) DNA methylation in early mammalian development. In: Razin H, Cedar H, Riggs AD (eds) DNA methylation, biochemistry and biological significance. Springer, Berlin Heidelberg New York, pp 189-219

Jones EC, Krohn PL (1961) The relationship between age, numbers of oocytes and fertility in virgin and multiparous mice. J Endocrinol 21: 469-494

Kang Y-H (1974) Development of the zona pellucida in the rat oocyte. Am J Anat 139: 535-566

Kaplan G, Abreu S, Bachvarova R (1982) rRNA accumulation and protein synthesis patterns in growing mouse oocytes. J Exp Zool 220: 361-370

Kaufman MH, Barton SD, Surani MAH (1977) Normal postimplantation development of mouse parthenogenetic embryos to the forelimb bud stage. Nature 265: 53-55

Konrad KD, Engstrom L, Perrimon N, Mahowald AP (1985) Genetic analysis of oogenesis and the role of maternal gene expression in early development. In: Browder L (ed) Developmental biology: A comprehensive synthesis, vol 1. Plenum, New York, pp 577-617

LaMarca MJ, Wassarman PM (1979) Ribosomal protein synthesis in mouse eggs and embryos. Dev Biol 73: 103-119

Levinson J, Goodfellow P, Vadeboncoeur M, McDevitt H (1978) Identification of stage-specific polypeptides synthesized during murine preimplantation development. Proc Natl Acad Sci USA 75: 3332-3336

McGrath J, Solter D (1984) Completion of mouse oogenesis requires both the maternal and paternal genomes. Cell 37: 179-183

Mahi-Brown CA, Huang TTF Jr, Yanagimachi R (1982) Infertility in bitches induced by active immunization with porcine zonae pellucidae. J Exp Zool 222: 89-95

Mangia F, Epstein CJ (1975) Biochemical studies of growing mouse oocytes: preparation of oocytes and analysis of glucose-6-phosphate dehydrogenase and lactate dehydrogenase activities. Dev Biol 45: 211-220

Mangia F, Erickson RP, Epstein C (1976) Synthesis of LDH-1 during mammalian oogenesis and early development. Dev Biol 54: 146-150

Mintz B, Russell ES (1957) Gene-induced embryological modifications of primordial germ cells in the mouse. J Exp Zool 134: 207-237
Ozdzenski W (1967) Observations on the origin of primordial germ cells in the mouse. J Embryol Exp Morphol 63: 75-84
Pedersen T (1969) Follicle growth in the immature mouse ovary. Acta Endocrinol (Copenh) 62: 117-132
Perona RM, Wassarman PM (1986) Mouse blastocysts hatch in vitro by using a trypsin-like proteinase associated with cells of mural trophectoderm. Dev Biol 113: 42-52
Peters H (1969) The development of the mouse ovary from birth to maturity. Acta Endocrinol (Copenh) 62: 98-116
Philpott CC, Ringuette MR, Dean J (1987) Oocyte-specific expression and developmental regulation of ZP-3, the sperm receptor of the mouse zona pellucida. Dev Biol 121: 568-575
Pratt HPM, Bolton VN, Gridgeon KA (1983) The legacy from the oocyte and its role in controlling early development of the mouse embryo. Ciba Found Symp 98: 197-227
Raff RA, Showman RM (1985) Maternal messenger RNA: quantitative, qualitative and spatial control of its expression in embryos. In: Metz CB, Monroy A (eds) Biology of fertilization, vol 3. Academic, Orlando, pp 401-452
Richter JO, McGaughey RW (1981) Patterns of polypeptide synthesis in mouse oocytes during germinal vesicle breakdown and during maintenance of the germinal vesicle stage by dibutyryl cAMP. Dev Biol 83: 188-192
Ringuette MJ, Sobieski DA, Chamow SM, Dean J (1986) Oocyte-specific gene expression: molecular characterization of a cDNA coding for ZP-3, the sperm receptor of the mouse zona pellucida. Proc Natl Acad Sci USA 83: 4341-4345
Sacco AG (1979) Inhibition of fertility in mice by passive immunization with antibodies to isolated zonae pellucida. J Reprod Fertil 56: 533-537
Saling PM, Sowinski J, Storey BT (1979) An ultrastructural study of epididymal mouse spermatozoa binding to zonae pellucidae in vitro: sequential relationship to acrosome reaction. J Exp Zool 209: 229-238
Salzmann GS, Greve JM, Roller RJ, Wassarman PM (1983) Biosynthesis of the sperm receptor during oogenesis in the mouse. EMBO J 2: 1451-1456
Sanford JP, Forrester L, Chapman VM (1984) Methylation patterns of repetitive DNA sequences in germ cells of Mus musculus. Nucleic Acid Res 12: 2823-2836
Sanford JP, Rossant J, Chapman VM (1985) DNA methylation patterns in germ cells and early embryos of the mouse. J Cell Biol 101: 456a
Schultz RM, Letourneau GE, Wassarman PM (1979) Program of early development in the mammal: changes in the pattern and absolute rates of tubulin and total protein synthesis during oocyte growth in the mouse. Dev Biol 73: 120-133
Shimizu S, Tsuji M, Dean J (1983) In vitro biosynthesis of three sulfated glycoproteins of murine zonae pellucidae by oocytes grown in follicle culture. J Biol Chem 258: 5858-5863
Smith LD, Richter JD (1985) Synthesis, accumulation and utilization of maternal macromolecules during oogenesis and oocyte maturation. In: Metz CB, Monroy A (eds) Biology of fertilization, vol 1. Academic, Orlando, pp 141-188
Soriano P, Jaenisch R (1986) Retroviruses as probes for mammalian development: allocation of cells to the somatic and germ cell lineages. Cell 46: 19-29
Surani MAH, Barton SC, Norris ML (1984) Development of reconstituted mouse eggs suggest imprinting of the genome during gametogenesis. Nature 308: 548-550
Surani MAH, Barton SC, Norris ML (1986) Nuclear transplantation in the mouse: heritable differences between parental genomes after activation of the embryonic genome. Cell 45: 127-136
Tam PPL, Snow MHL (1981) Proliferation and migration of primordial germ cells during compensatory growth in mouse embryos. J Embryol Exp Morphol 64: 133-147
Tsunoda Y, Whittington DG (1982) Lack of effect of zona antibody on the development of mouse embryos in vivo and in vitro. J Reprod Fertil 66: 585-589
Tsunoda Y, Sugie T, Mori J, Isojima S, Koyama K (1981) Effect of purified zona antibody on fertilization in the mouse. J Exp Zool 217: 103-108
Van Blerkom J (1981) Structural relationship and post translational modification of stage-specific proteins synthesized during early preimplantation development in the mouse. Proc Natl Acad Sci USA 78: 7629-7633

/assarman PM (1987) The biology and biochemistry of fertilization. Science 235: 553-560

/assarman PM, Josefowicz WJ (1978) Oocyte development in the mouse: an ultrastructural comparison of oocytes isolated at various stages of growth and meiotic competence. J Morphol 156: 209-236

Volf DP (1981) The mammalian egg's block to polyspermy. In: Mastroianni L Jr, Biggers JD (eds) Mechanisms of fertilization and embryonic development in vitro. Plenum, New York, pp 183-197

Volgemuth DJ, Celenza J, Bundman DS, Dunbar BS (1984) Formation of the rabbit zona pellucida and its relationship to ovarian follicular development. Dev Biol 106: 1-14

Vood DM, Liu C, Dunbar BS (1981) Effect of alloimmunization and heteroimmunization with zonae pellucidae on fertility in rabbits. Biol Reprod 25: 439-450

Vylie CC, Stott D, Donovan PJ (1985) Primordial germ cell migration. In: Browder L (ed) The cellular basis of morphogenesis. Plenum, New York (Developmental biology: A comprehensive synthesis, vol 2)

'anagimachi R (1981) Mechanisms of fertilization in mammals. In: Mastroianni L Jr, Biggers JD (eds) Mechanisms of fertilization and embryonic development in vitro. Plenum, New York, pp 81-182

The Mouse Egg Receptor For Sperm: A Multifunctional Zona Pellucida Glycoprotein*

P. Wassarman, J. Bleil, C. Fimiani, H. Florman, J. Greve, R. Kinloch, C. Moller, S. Mortillo, R. Roller, G. Salzmann, and M. Vazquez

Introduction

Fertilization is the process by which sperm and eggs unite to form a zygote, the true beginning of a new individual. Among mammals, sperm first make contact with eggs at the surface of the egg extracellular coat, or zona pellucida (ZP). This contact can lead to species-specific binding of sperm to eggs via sperm receptors located in the ZP. Bound sperm then undergo changes that enable them to penetrate the ZP and fuse with egg plasma membrane. Fusion results in activation of eggs, and development of the organism ensues [1-5].

Experiments carried out in our laboratory during the past 10 years or so have resulted in identification and isolation of the mouse egg sperm receptor, as well as in characterization of certain of its molecular and physiological properties [6-11]. Here, we summarize some salient features of the mouse sperm receptor, a glycoprotein called ZP3, regarding its biosynthesis, structure, and multiple functions (i.e., sperm receptor, acrosome reaction inducer, and ZP structural element).

Early Events in Gamete Interaction

To study the sperm receptor, we examined early events of mouse gamete interaction in vitro [5] (Fig. 1). Ovulated eggs and two-cell embryos were added to capacitated sperm and within seconds ZP of both eggs and embryos were covered with motile sperm. These sperm were loosely associated with ZP and could be removed by gentle pipetting with a broad-bore micropipet; this state of adhesion is referred to as "attachment." Shortly thereafter, contact between sperm and egg ZP became more tenacious, such that gentle pipetting no longer dissociated the gametes; this state of adhesion is referred to as "binding." Although the initial reversible attachment of sperm to embryo ZP was virtually indistinguishable from that observed with eggs, in the former case attachment did not proceed to the binding state. This pronounced difference in behavior provides an operational definition of bound sperm as those adhering to egg ZP under conditions that result in complete removal of sperm from embryo ZP (Fig. 2).

---

* Research in our laboratory was supported in part by the National Institute of Child Health and Human Development, National Science Foundation, Rockefeller Foundation's and Hoffman - La Roche Inc.

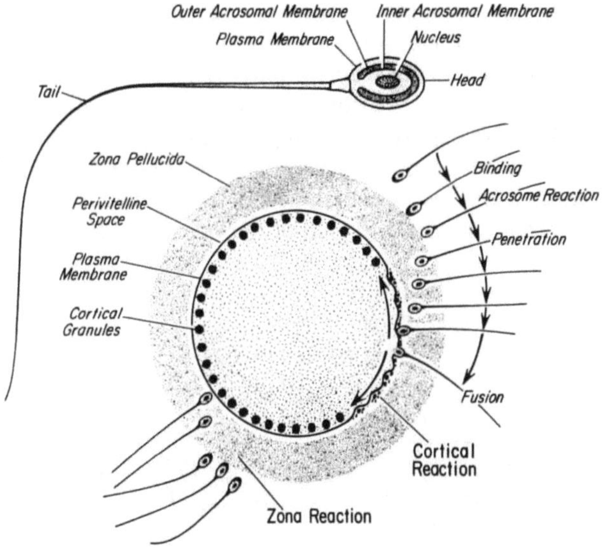

Fig. 1. Diagrammatic representation of mouse gametes and the fertilization pathway in mice. The sequence of events includes attachment of sperm to eggs, followed by binding of sperm to eggs, completion of the acrosome reaction, penetration of the ZP, sperm-egg fusion, cortical reaction, and zona reaction. (From [4])

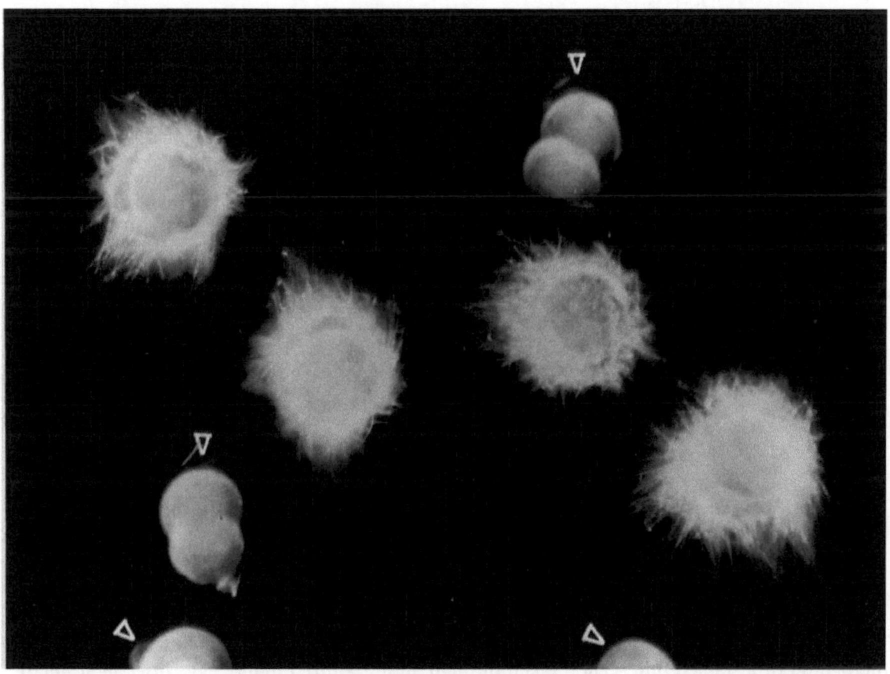

Fig. 2. Light micrograph illustrating the binding of mouse sperm to ZP of unfertilized mouse eggs, but not to ZP of two-cell mouse embryos in vitro. *Arrowheads* indicate two-cell embryos

In mice, binding of sperm to egg ZP occurs at the plasma membrane overlying the anterior region of the sperm head. Shortly after binding, sperm undergo the so-called acrosome reaction (AR). This is a membrane fusion event, wherein the contents of the acrosome, a large lysosome-like vesicle overlying the sperm nucleus, are exposed following vesiculation of the anterior region of sperm plasma membrane and outer acrosomal membrane (Fig. 3). Completion of the AR is required for sperm penetration of ZP as well as for fusion of plasma membrane toward the posterior region of the sperm head with egg plasma membrane.

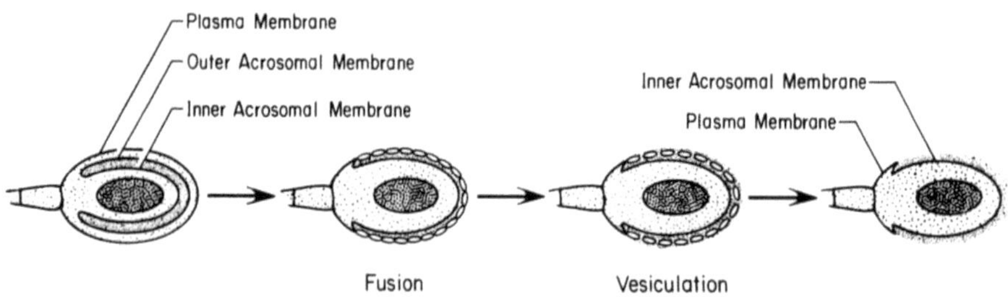

Fig. 3. Diagrammatic representation of mouse sperm undergoing the acrosome reaction. The process involves fusion of plasma and outer acrosomal membrane at many sites along the anterior portion of the sperm head, formation of hybrid membrane vesicles, and exposure of inner acrosomal membrane with associated acrosomal contents. (From [4])

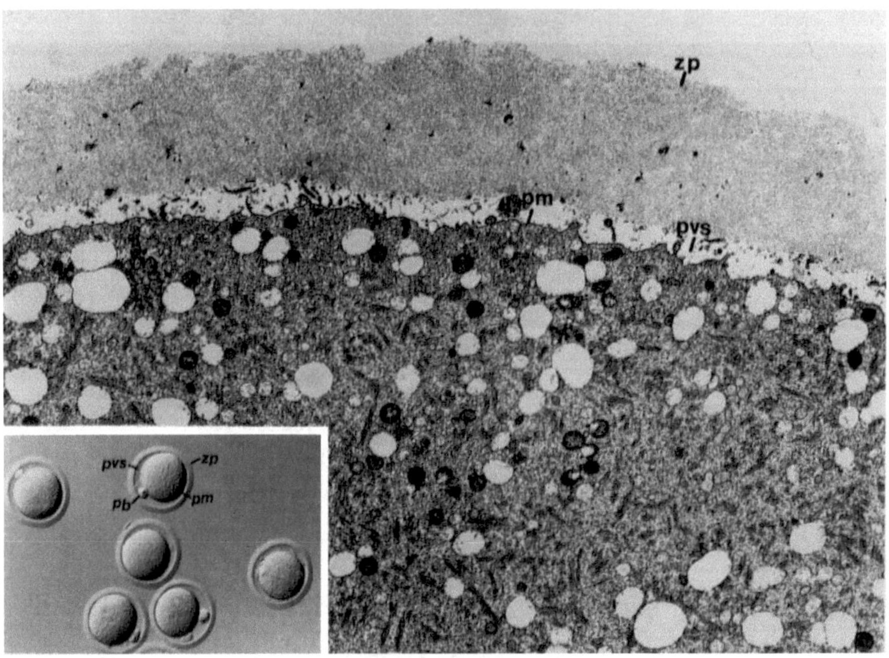

Fig. 4. Light- and electron-microscopic views of unfertilized mouse eggs. Shown is an electron micrograph of a stained thin section taken through an unfertilized mouse egg. *Inset,* light micrograph of a group of unfertilized mouse eggs. *zp,* Zona pellucida; *pvs,* perivitelline space; *pm,* plasma membrane; *pb,* polar body

Nature of the Zona Pellucida

The ZP is a relatively thick extracellular coat that surrounds all mammalian eggs and plays important roles during both fertilization and preimplantation development [1-5]. For example, ZP contain species-specific sperm receptors that function during fertilization, participate in a secondary block to polyspermy following fertilization, and protect early embryos as they pass down the oviduct into the uterus. It is not until the expanded blastocyst stage of development that embryos hatch from their ZP, thereby enabling them to implant in the uterine wall.

Mouse egg ZP are about 7 μm thick and contain about 3 ng protein [1-5]. Transmission electron micrographs of sectioned material indicate that the ZP is an amorphous network, not unlike a number of other types of extracellular coats in appearance (Fig. 4; discussed below). As such, ZP do not present a permeability barrier to large macromolecules or even to small viruses. ZP are dissolved by a variety of agents that either do or do not disrupt covalent bonds, with egg ZP being more susceptible to these agents than embryo ZP. The latter state is thought to reflect a "hardening" of ZP following fertilization of eggs.

Glycoproteins of the Zona Pellucida

The 3 ng or so of egg ZP protein is distributed among three glycoprotein species, called ZP1, ZP2, and ZP3, having apparent M_r of 200000, 120000 and 83000, respectively, by sodium dodecyl sulfate polycrylamide gel electrophoresis (SDS-PAGE; Bleil and Wassarman 1980; Wassarman 1988; see Fig. 5). These three glycoproteins together account for virtually all the mass of isolated, thoroughly washed ZP. ZP2 and ZP3 are present in nearly equimolar amounts, accounting for as much as 80% of ZP protein. Although the glycoproteins exhibit considerable heterogeneity on SDS-PAGE, a variety of evidence suggests that each glycoprotein represents a unique polypeptide chain and that glycosylation gives rise to the observed heterogeneity (discussed below). All three ZP glycoproteins are acidic ($pI < 5.5$), in part due to the presence of terminal sialic acid on their oligosaccharides, and only ZP1 consists of more than one polypeptide chain (linked by intermolecular disulfides). These are the mature forms of the glycoproteins secreted by oocytes (discussed below) and found in the ZP.

Is the relatively simple glycoprotein composition of the ZP compatible with its known properties and functions? In principle, yes. However, it requires that ZP glycoproteins have domains involved in assembly and maintenance of structure of the insoluble coat ("structural domains"), as well as domains involved in spermegg interaction during fertilization ("functional domains"). Such requirements are not unlike those known to be satisfied by macromolecular components of other types of extracellular matrices. Thus far, analyses of mouse ZP glycoproteins suggest that they too satisfy these requirements (discussed below). In this context, it is interesting to note that, whereas mammalian eggs have a single extracellular coat, the ZP nonmammalian eggs frequently have two coats - a relatively thin vitelline envelope surrounding the plasma membrane and an outer, thick jelly coat. Among mammals, the ZP must serve the combined functions performed by the vitelline envelope and jelly coat during fertilization in nonmammals.

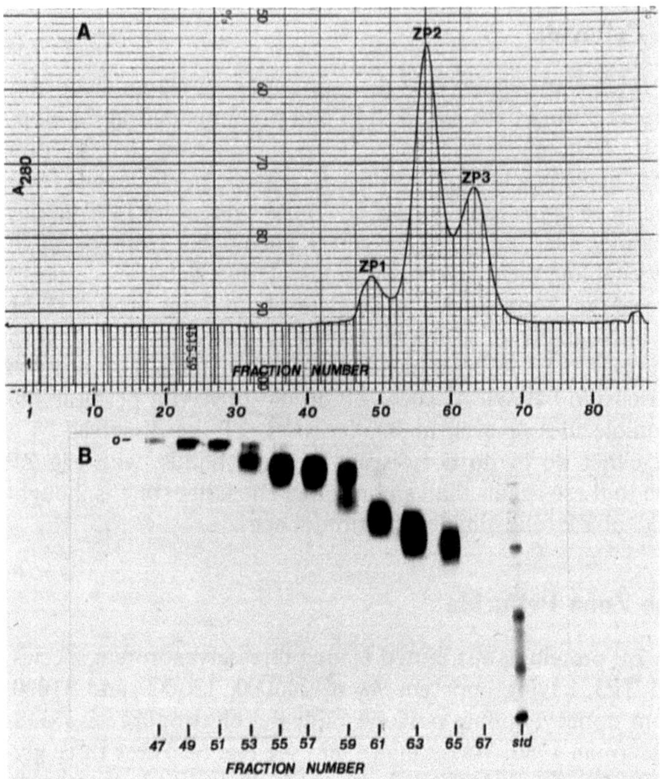

Fig. 5 A, B. HPLC purification of ZP glycoproteins. Purified ZP from mouse ovaries [see ref. 18] were solubilized and subjected to HPLC. **A** A profile of absorption at 280 nm for a preparation of approximately 30000 ZP (positions of ZP1, ZP2, and ZP3 are indicated). **B** Fractions were collected and aliquots were subjected to SDS-PAGE followed by staining with silver. Note the fractions containing purified ZP1 (49-51), ZP2 (53-57), or ZP3 (61-65)

Identification of the Sperm Receptor

Based upon the fertilization pathway in mice [described above; refs. 4,5], a sperm receptor purified from ZP would be expected to satisfy certain requirements in vitro:

1. Receptor purified from egg ZP should bind to sperm and, consequently, prevent their binding to ovulated eggs (i.e., prevent fertilization).
2. Receptor purified from embryo ZP should not bind to sperm and not prevent their binding to ovulated eggs (i.e., not prevent fertilization).
3. Receptor purified from egg ZP should bind to heads of acrosome-intact sperm, but not to heads of sperm that have undergone the AR.
4. Receptor purified from egg ZP should not bind to a wide variety of cell types.
5. Receptor purified from egg ZP should exhibit a certain degree of species specificity in binding to sperm.

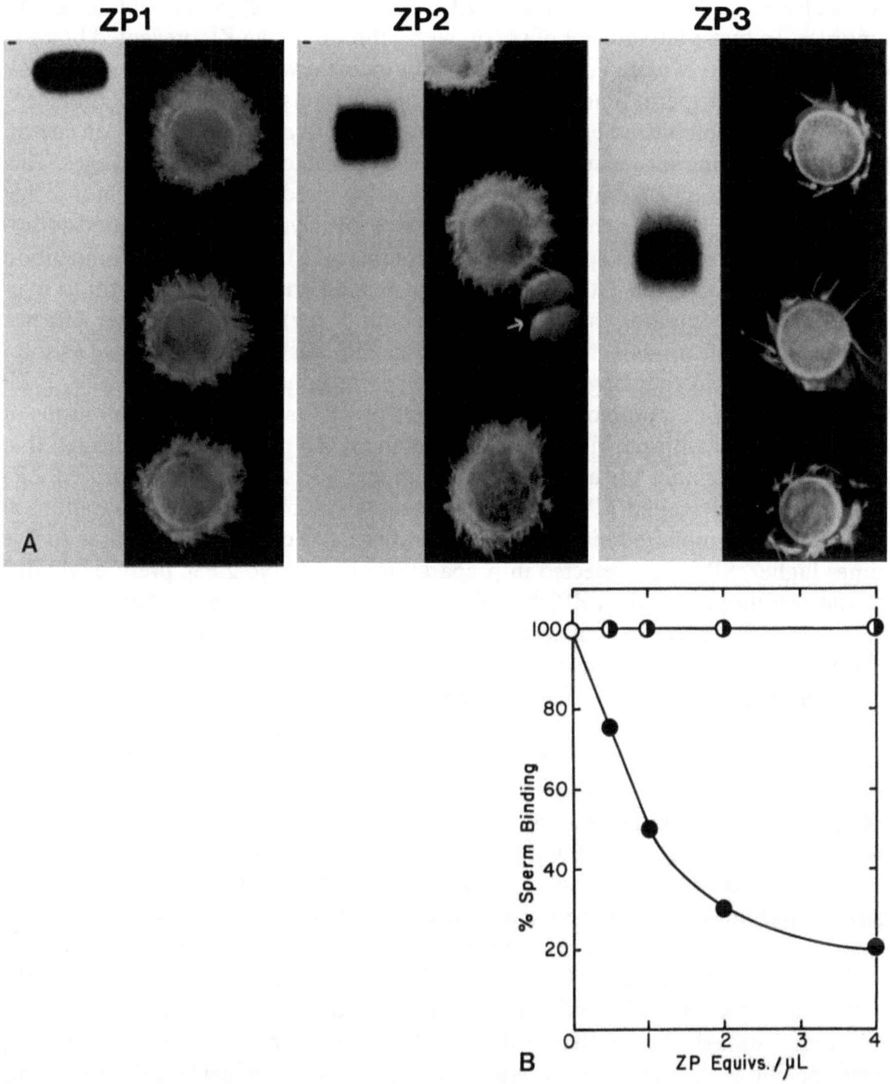

Fig. 6. **A** Effect of individual ZP glycoproteins on the binding of mouse sperm to eggs in vitro. Shown are autoradiograms of radiolabeled ZP1, ZP2, and ZP3, purified from egg ZP, following SDS-PAGE on nonreducing gels and photomicrographs of sperm bound to eggs after exposure of sperm to ZP1, ZP2, or ZP3. **B** Quantitation of the effect of individual ZP glycoproteins on the binding of mouse sperm to eggs in vitro. Binding of sperm to eggs was compared after exposure of sperm to ZP1 *(half-filled circles)*, ZP2 *(half-filled circles)*, or ZP3 *(filled circles)* purified from egg ZP, or to culture medium alone *(open circle)*. The number of sperm bound to eggs was determined as a function of ZP glycoprotein concentration (expressed as ZP equivalents per microliter). (From [8])

As indicated above, sperm bind to ZP of unfertilized eggs but not to ZP of embryos. In fact, pretreatment of sperm with solubilized egg ZP prevented binding of these sperm to ovulated eggs, whereas pretreatment with solubilized embryo ZP had no effect on sperm binding [14]. These observations suggest that sperm receptors present in solubilized egg ZP preparations bind to appropriate sites on sperm, thereby preventing subsequent binding of these sperm to ZP of ovulated eggs. This formed the basis of a "competition assay" that was used to identify the mouse egg sperm receptor [8,14]. Each of the three mouse ZP glycoproteins, purified to homogeneity, was tested for sperm receptor activity (SRA) in the in vitro competition assay. In each case, the extent of binding of treated and untreated sperm to ovulated eggs was compared (number of sperm bound/egg). Whereas neither ZP1 nor ZP2 had a significant effect on sperm binding, ZP3 was nearly as effective as solubilized egg ZP in reducing the number of sperm bound to eggs ($ID_{50} \simeq 1$ ZP equivalent/µl; i.e., nanomolar concentrations of ZP3; see Fig. 6). Comparisons of results obtained with purified ZP3 and solubilized ZP preparations indicated that ZP3 alone accounted for all SRA present in egg ZP. As expected, purified ZP3 from embryo ZP (called $ZP3_f$) had no effect on sperm binding at concentrations at which egg ZP3 inhibited binding by 80% or more. (At concentrations three to five times higher, SRA was detected in preparations of embryo $ZP3_f$, presumably due to the presence of residual ZP3 [8]). These results strongly suggest that ZP3 serves as receptor for sperm in ovulated eggs, and that ZP3 is modified as a result of fertilization such that it no longer exhibits SRA.

Mouse sperm that have undergone the AR fail to bind to unfertilized eggs; only acrosome-intact sperm bind [5,15-17]. In this context, autoradiograms of whole-mount preparations of sperm incubated with ^{125}I-radiolabeled ZP3 revealed that ZP3 was bound exclusively to heads of acrosome-intact sperm, not to heads of sperm that had undergone the AR [18] (Fig. 7). Although both red blood cells and residual bodies were present in these sperm preparations, no silver grains due to radiolabeled ZP3 were found associated with these contaminants. Under the same conditions, radiolabeled fetuin was bound at only background levels to heads of both acrosome-intact sperm and sperm that had undergone the AR, and radiolabeled ZP2 bound preferentially to heads of sperm that had undergone the AR. Therefore, while ZP3 and fetuin share some common molecular features (e.g., N- and O-linked oligosaccharides), only ZP3 was bound to sperm, and binding was specific with respect to cell type (i.e., only sperm), cellular localization (only the sperm head), and cellular state (only acrosome-intact sperm). Furthermore, although ZP2 and ZP3 also share some common molecular features, they recognize different antigenic determinants on the sperm head. In the former case, determinants associated with inner acrosomal membrane and in the latter, determinants associated with plasma membrane. Quantitation of autoradiographic data indicated that saturation of ZP3 binding sites on sperm (called "egg-binding protein") occurred in the range $10-50 \times 10^3$ ZP3 molecules bound per sperm head. (Results of calculations suggest that the mouse sperm head can accomodate this number of receptor molecules, given the approximate dimensions of the ZP2: ZP3 dimers – $\simeq 7 \times 14$ nm – that constitute the structural repeat of mouse ZP filaments; discussed below.)

Finally, mouse sperm will also bind to ZP3 covalently linked to silica beads

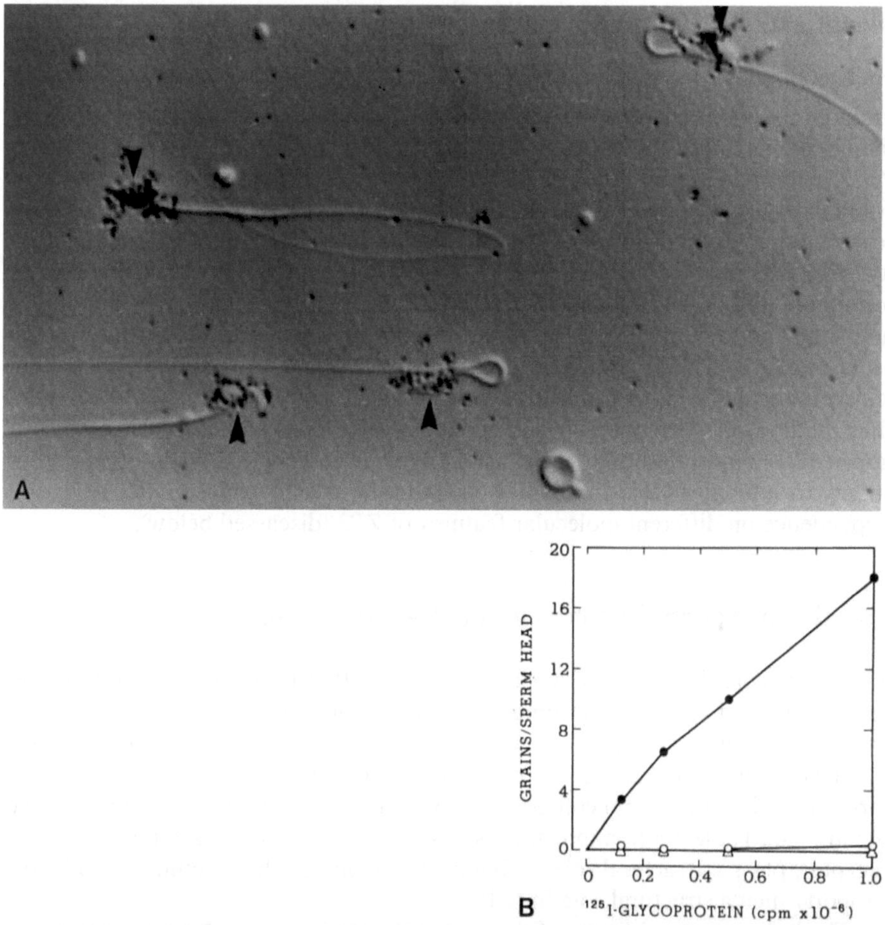

Fig. 7. **A** Autoradiographic visualization of radiolabeled mouse sperm receptor (ZP3) bound to acrosome-intact mouse sperm. Arrowheads indicate silver grains localized to the sperm head. **B** Quantitation of binding of radiolabeled ZP3 to acrosome-intact and acrosome-reacted sperm. Sperm were incubated in the presence of ^{125}I- labeled ZP3 *(filled and open circles)* or ^{125}I-labeled fetuin *(triangles)*, subjected to whole-mount autoradiography and visualized by Nomarski DIC optics. Shown are the average number of silver grains per sperm head, plotted as a function of either ^{125}I- labeled ZP3 or fetuin concentration, for both acrosome-intact *(filled circles, triangles)* and acrosome-reacted *(open circles, triangles)* sperm. (From [18])

[19; M. Vazquez and P. Wassarman, unpublished results]. (Sperm do not bind to unmodified silica beads or to beads to which bovine serum albumin, fetuin, or ZP2 has been covalently linked, indicating binding specificity for ZP3.) As in the case of unfertilized eggs, only acrosome-intact sperm were bound to ZP3 beads, and sperm pretreated with egg ZP3 failed to bind to the beads. As expected, sperm did not bind to beads to which $ZP3_f$ had been covalently linked.

Identification of the Acrosome Reaction Inducer

Once bound to the ZP of ovulated eggs, acrosome-intact mouse sperm must undergo the AR in order to be able to penetrate the ZP and fuse with egg plasma membrane [5]. Since the AR occurs on the ZP, it was not surprising to find that a ZP glycoprotein induced the AR in vitro [17]. Sperm incubated in the presence of solubilized egg ZP underwent the AR, whereas solubilized embryo ZP was ineffective. When each of the egg ZP glycoproteins was tested for AR-inducing activity, only ZP3 induced sperm to undergo the AR in vitro; ZP3 was as effective as ionophore A23187 at inducing the reaction. These results suggest that binding of acrosome-intact mouse sperm to ZP3, through plasma membrane overlying the anterior region of the sperm head, is sufficient to alter the affected plasma so that it becomes capable of fusing with outer acrosomal membrane. It should be noted that binding of sperm to ZP3 and induction of the AR by ZP3 can be distinguished from one another by their distinctive time courses [17,19], differential sensitivity to inhibitors [16], differential dependence on ZP3 concentration [17], and dependence on different molecular features of ZP3 (discussed below).

Location of Sperm Receptors in the Zona Pellucida

There are more than a billion copies of ZP3 in ZP of fully grown mouse oocytes and unfertilized eggs. ZP3 accumulates in ZP of growing oocytes as each oocyte increases in diameter (from $\simeq 12$ to $\simeq 85$ μm), and its ZP increases in thickness [13; discussed below]. Therefore, even small, growing oocytes have sperm receptors in their ZP, and sperm consequently bind to isolated growing oocytes cultured in vitro [14]. These observations suggest that ZP3, in addition to serving as a sperm receptor, plays a structural role during ZP assembly. Evidence summarized below supports such a structural role for ZP3.

The basic structural theme of the mouse ZP has been revealed by transmission electron microscopy (EM) of solubilized ZP preparations, either sprayed as a fine mist onto a substrate or adsorbed onto Formvar-coated grids, as well as by chemical cross-linking and antibody decoration studies of solubilized preparations [13,20; J. Greve and P. Wassarman, unpublished results]. Two of the solubilization procedures used, elastase and pH 5.5 treatment of isolated ZP, produced primarily large complexes consisting of interconnected or branched filaments in which heterologous interactions between ZP1, ZP2, and ZP3 were maintained. Two alternative solubilization procedures, chymotrypsin and dithiothreitol treatment of isolated ZP, yielded primarily individual, or unbranched filaments in which heterologous interactions, at least between ZP2 and ZP3, were maintained. Our observations, summarized below, provide a low-resolution view of mouse ZP structure.

The ZP is composed of long (2–3 μm in solubilized preparations) interconnected filaments of uniform width ($\simeq 7$ nm by negative staining; $\simeq 18$ nm by rotary shadowing; Fig. 8).

ZP filaments exhibited a structural repeat of about 14–15 nm in EM. The appearance of the repeat depended upon the way in which filaments were pre-

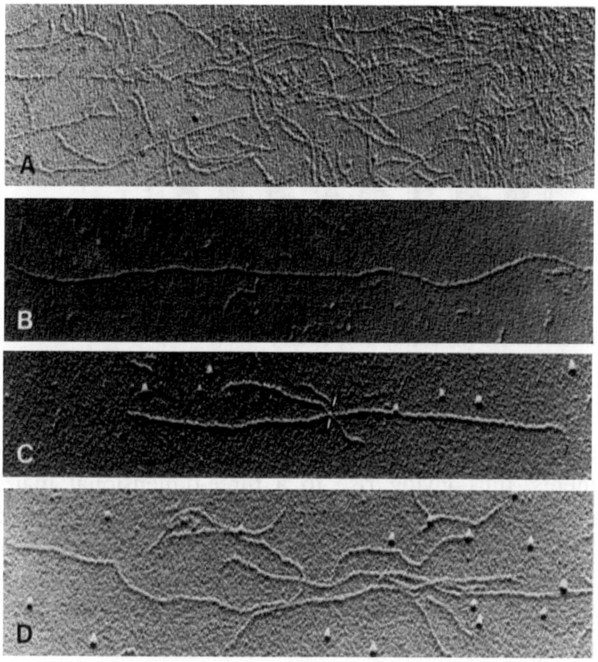

Fig. 8. Electron micrographs of mouse ZP filaments, sprayed onto mica-coated grids, and platinum/carbon shadowed. Filaments were prepared by solubilizing isolated ZP with elastase (see [20]). The filament shown in *panel B* is more than 2 μm in length. Note the interconnection between filaments in *panel C* (indicated by *white bars*)

pared for EM. For example, the repeat was somewhat obscured in sprayed and shadowed preparations but seen clearly in negatively stained preparations adsorbed (not sprayed) onto grids, as well as in rapidly frozen, shadowed preparations. A repeat was also seen in sprayed and negatively stained preparations; however, the presence of glycerol in samples resulted in collapse of the repeat into "beads" (about 9 nm in diameter; about 17 nm repeat) along ZP filaments.

The structural repeat observed with ZP filaments reflects the presence of a ZP2:ZP3 heterodimer every 14 nm or so along filaments. This was determined by SDS-PAGE of chemically cross-linked isolated filaments and intact ZP, as well as by EM of filaments decorated with monoclonal antibodies directed against either ZP2 or ZP3. The former approach resulted in identification of an approximately 180000 M_r ZP2:ZP3 dimer, as well as higher M_r oligomers, and the latter approach revealed the presence of ZP2 and ZP3 every 14 nm or so along ZP filaments. These and other observations suggest that the ZP is composed of filaments that are polymers constructed of ZP2: ZP3 dimers. Noncovalent bonds stabilize ZP2: ZP3 dimers, as well as interactions between dimers (Fig. 9).

Treatment of intact ZP with dithiothreitol resulted in reduction of the intermolecular disulfides of ZP1, whereas treatment with chymotrypsin resulted in limited proteolysis of ZP1. In both cases, ZP were dissolved and the solubilized preparations consisted of filaments that were no longer interconnected. These observa-

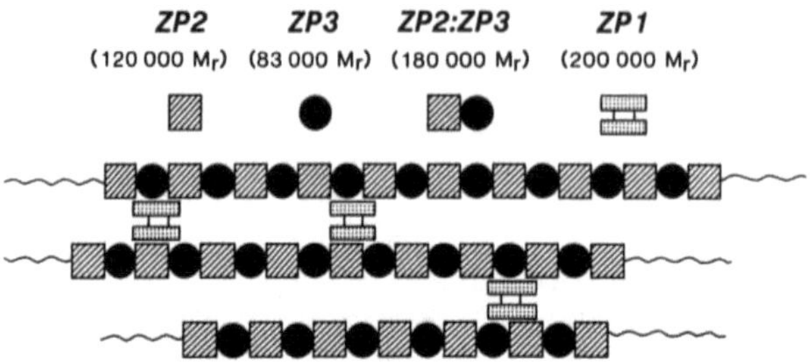

Fig. 9. Diagrammatic representation of the arrangement of glycoproteins in mouse ZP filaments. This model features polymerization of ZP2-ZP3 dimers into long ZP filaments, with filament interconnections attributable to ZP1

tions suggest that ZP1 is responsible for making interconnections between ZP filaments, thus giving rise to a three-dimensional matrix. Since ZP1 is a dimer composed of identical polypeptide chains, the molecule could potentially interact with two different filaments. Thus far, it is unclear whether ZP1 is a ZP filament "associated protein" or an integral component of filaments, although the former possibility seems most likely based on available evidence.

In conclusion, although only ZP3 molecules located at the ZP surface participate in initial binding of sperm to eggs (i.e., serve as sperm receptors), ZP3 molecules throughout the ZP participate in maintaining structural integrity of the extracellular coat. This explains, at least in part, the presence of ZP3 in ZP of small, growing oocytes that are incapable of being fertilized. The dual role of ZP3, as receptor and structural element, is not unique in the context of extracellular matrix components of somatic cells.

Structure and Expression of the Sperm Receptor

We found that isolated growing oocytes, denuded of follicle cells and cultured in vitro in the presence of [^{35}S]methionine or [^{3}H]fucose, synthesized and secreted all three ZP glycoproteins [8,13,21-24]. Accordingly, pulse-chase experiments utilizing either enzymes or chemical agents that remove N- and/or O-linked oligosaccharides from glycoproteins, and experiments carried out in the presence of tunicamycin to prevent N-linked glycosylation of nascent proteins, permitted construction of a biosynthetic pathway for ZP3 (Fig. 10). The glycoprotein is synthesized as a 44000 M_r polypeptide chain (discussed below) to which either three or four high mannose-type, N-linked oligosaccharides are added cotranslationally, giving rise to 53000 and 56000 M_r intermediates, respectively. The N-linked oligosaccharides are subsequently processed to complex-type in the Golgi, and O-linked oligosaccharides are added to nascent ZP3 (83000 M_r) prior to its secretion from oocytes.

Recently, a cDNA encoding a 20 amino acid stretch of ZP3 polypeptide chain was isolated and partially characterized [25]. Based on the reported sequence of the ZP3 cDNA fragment, we prepared a synthetic oligonucleotide probe, used it

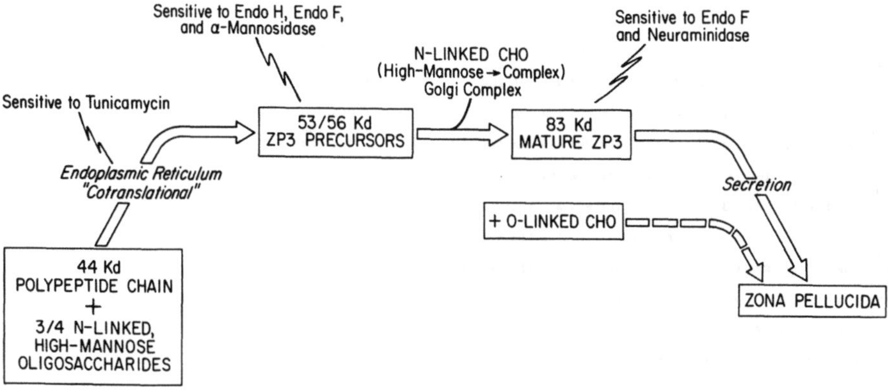

Fig. 10. Biosynthetic pathway for ZP3 in growing mouse oocytes. (From [8, 23, 24])

to screen CD-1 mouse genomic libraries, and partially characterized two overlapping clones that span approximately 10 kb of the ZP3 locus. These genomic clones encompass the entire ZP3 coding region, which contains eight exons. The exons were identified, mapped, and sequenced, yielding the entire primary structure of the polypeptide chain [26; R. Kinloch and P. Wassarman, unpublished results]. ZP3 DNA and polypeptide sequences were used to search Genbank and PIR data-bases, respectively. Significant homologies with data-base entries were not found in either case.

The predicted primary structure of the ZP3 polypeptide chain consists of 424 amino acids, with the N-terminal 22 amino acids predicted to comprise a signal sequence (Fig. 11). The latter has a short polar domain followed by a hydrophobic region consisting of 11 amino acids. The 4 amino acids around the cleavage site at residue 22 of the signal sequence comply with the -3, -1 rule for such sequences [27]. Processing of the signal sequence leaves a polypeptide chain consisting of 402 amino acids (43900 M_r). This is in good agreement with previous assessments of M_r of ZP3 (44000) based on SDS-PAGE analysis of intracellular precursors metabolically radiolabeled in the presence of tunicamycin [23,24]. Furthermore, the sequence predicted for residues 184–200 is identical to that determined directly for a ZP3 peptide produced by digestion of the glycoprotein with V8 protease (R. Roller and P. Wassarman, unpublished results). The N terminus of the ZP3 polypeptide chain is blocked, suggesting that, following cleavage of the signal sequence, the N-terminal glutamine cyclizes to form a pyroglutamyl residue [28].

As compared to the average vertebrate protein, ZP3 contains an unusually high number of histidine, valine, serine, threonine, and proline residues, and low number of methionine, tyrosine, lysine, and isoleucine residues. Serine plus threonine (72 residues) and proline (30 residues) represent approximately 17% and 7%, respectively, of the total amino acid composition. The high serine plus threonine content of ZP3 is not unexpected given the relatively high degree of glycosylation of the polypeptide chain. It is of interest to note that the middle third of the polypeptide chain (residues 119–259) includes 17 of the 30 proline residues, with two proline-rich domains (residues 119–159, 7 prolines; residues 208–259, 9 prolines)

MET ALA SER SER TYR PHE LEU PHE LEU CYS LEU LEU LEU CYS GLY GLY PRO GLU LEU CYS
1 20
ASN SER GLN THR LEU TRP LEU LEU PRO GLY GLY THR PRO THR PRO VAL GLY SER SER SER
21 40
PRO VAL LYS VAL GLU CYS LEU GLU ALA GLU LEU VAL VAL THR VAL SER ARG ASP LEU PHE
41 60
GLY THR GLY LYS LEU VAL GLN PRO GLY ASP LEU THR LEU GLY SER GLU GLY CYS GLN PRO
61 80
ARG VAL SER VAL ASP THR ASP VAL VAL ARG PHE ASN ALA GLN LEU HIS GLU CYS SER SER
81 100
ARG VAL GLN MET THR LYS ASP ALA LEU VAL TYR SER THR PHE LEU LEU HIS ASP PRO ARG
101 120
PRO VAL SER GLY LEU SER ILE LEU ARG THR ASN ARG VAL GLU VAL PRO ILE GLU CYS ARG
121 140
TYR PRO ARG GLN GLY ASN VAL SER SER HIS PRO ILE GLU PRO THR TRP VAL PRO PHE ARG
141 160
ALA THR VAL SER SER GLU GLU LYS LEU ALA PHE SER LEU ARG LEU MET GLU GLU ASN TRP
161 180
ASN THR GLU LYS SER ALA PRO THR PHE HIS LEU GLY GLU VAL ALA HIS LEU GLN ALA GLU
181 200
VAL GLN THR GLY SER HIS LEU PRO LEU GLN LEU PHE VAL ASP HIS CYS VAL ALA THR PRO
201 220
SER PRO LEU PRO ASP PRO ASN SER SER PRO TYR HIS PHE ILE VAL ASP PHE HIS GLY CYS
221 240
LEU VAL ASP GLY LEU SER GLU SER PHE SER ALA PHE GLU VAL PRO ARG PRO ARG PRO GLU
241 260
THR LEU GLN PHE THR VAL ASP VAL PHE HIS PHE ALA ASN SER SER ARG ASN THR LEU TYR
261 280
ILE THR CYS HIS LEU LYS VAL ALA PRO ALA ASN GLN ILE PRO ASP LYS LEU ASN LYS ALA
281 300
CYS SER PHE ASN LYS THR SER GLN SER TRP LEU PRO VAL GLU GLY ASP ALA ASP ILE CYS
301 320
ASP CYS CYS SER HIS GLY ASN CYS SER ASN SER SER SER SER GLN PHE GLN ILE HIS GLY
321 340
PRO ARG GLN TRP SER LYS LEU VAL SER ARG ASN ARG ARG HIS VAL THR ASP GLU ALA ASP
341 360
VAL THR VAL GLY PRO LEU ILE PHE LEU GLY LYS ALA ASN ASP GLN THR VAL GLU GLY TRP
361 380
THR ALA SER ALA GLN THR SER VAL ALA LEU GLY LEU GLY LEU ALA THR VAL ALA PHE LEU
381 400
THR LEU ALA ALA ILE VAL LEU ALA VAL THR ARG LYS CYS HIS SER SER SER TYR LEU VAL
401 420
SER LEU PRO GLN
421 424

Fig. 11. Primary structure of the ZP3 polypeptide chain predicted from exon sequences. The entire sequence of the 424 amino acid polypeptide chain is shown. The predicted 22 amino acid signal sequence is *italicized* and locations of six potential N-linked glycosylation sites (Asn-X-Ser/Thr) are *underlined*. (From [26])

separated by a proline-poor domain (159–207, 1 proline). The C-terminal 165 amino acids (residues 260–424) includes only 6 proline residues.

Examination of the ZP3 primary structure reveals six potential glycosylation sites for N-linked oligosaccharides, four of which are clustered within 60 amino acids of one another in the C-terminal half of the polypeptide chain. The sequence Asn-X-Ser/Thr, where X may be any amino acid (except, possibly Asp), is necessary for N-linked glycosylation and is always glycosylated in vitro when presented as a peptide or part of a denatured protein [29]. However, the sequence is not always used in vivo, perhaps, because it is not always accessible to the oligosaccharide transferase as nascent protein folds. Mature ZP3 possesses three or four complex-type, N-linked oligosaccharides [23]. Therefore, the six acceptor sequences are sufficient to account for the N-linked oligosaccharides found on ZP3. The extreme closeness of acceptor asparagine residues at positions 327 and 330 may mitigate against the use of both of them due to steric considerations.

Two computerized methods were used to analyze the available primary structure of ZP3, with the aim of identifying regions of sequence with a propensity to participate in secondary structures, such as α-helices, extended chains, reverse turns, and coils. Both analyses indicated that ZP3 sequence has very little α-helix forming potential. Rather, they suggested that the sequence consists of regions of

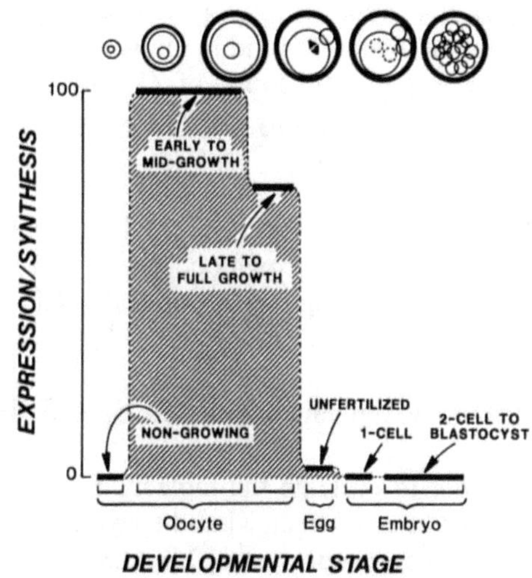

Fig. 12. Diagrammatic representation of ZP3 gene expression and ZP glycoprotein synthesis during oogenesis and early embryogenesis in the mouse. The diagram is based on measurements of levels of ZP3 mRNA (RNase protection experiments; R. Roller, R. Kinlach and P. Wassarman, unpublished results) and levels of ZP1, ZP2, and ZP3 synthesis ([22-24]; G. Salzmann and P. Wassarman, unpublished results). Undetectable levels of mRNA and synthesis are designated as 0 and maximal levels as 100

5-29 residues in an extended chain conformation, interupted by short, 4- to 10-residue, stretches of reverse turns or coils. Determination of local hydrophobicity along the sequence indicates that the sequence is neither hydrophobic nor hydrophilic, although two short hydrophobic domains are present at the C terminus (residues 363-369 and 385-407), in addition to the signal sequence hydrophobic domain at the N terminus of the polypeptide chain. An interesting aspect of these analyses is the finding that all six potential N-linked glycosylation sites are located in relatively hydrophilic stretches of the sequence that are predicted to form reverse turns or coils. In this connection, analysis of predicted secondary structures of glycoproteins having N-linked oligosaccharides, suggests that the majority of acceptor sequences utilized are present in β-turners or loops [29].

Finally, probes transcribed from restriction fragments of our ZP3 genomic clones have been used to follow ZP3 gene expression by an RNase protection assay (R. Roller, R. Kinloch, and P. Wassarman, unpublished results) (Fig. 12). Results of these experiments may be summarized as follows:

1. ZP3 mRNA ($\simeq$1.7 kb) was detected in total RNA prepared from mouse oocytes and ovaries, but not in total RNA from brain, heart, intestine, kidney, liver, muscle, testis, uterus, and 13-day embryos. The amount of ZP3 mRNA in ovaries was accounted for by oocyte ZP3 mRNA alone (confirmed by in situ hybridization) and was not detected in follicle cells.
2. ZP3 mRNA was found in growing and fully grown oocytes and accumulated during oocyte growth to a level of approximately 250000 copies per fully grown oocyte ($\simeq$220 fg/cell; $\simeq$1.8% of "translated" mRNA). Note: the level of ZP3

mRNA found in fully grown (≃85 μm in diameter) oocytes is approximately 70% of the level found in midgrowth stage (≃60–70 μm in diameter) oocytes.
3. Fewer than 5000 copies of ZP3 mRNA were found in ovulated eggs (≃50-fold reduction from fully grown oocyte levels), and ZP3 mRNA was not detected in two-cell embryos.

These results support our previous conclusions that ZP3 is expressed only in growing and fully grown oocytes (2- to 3-week period) and represents a major synthetic effort during this period [8,13,21,23,24].

Role of Oligosaccharides in Sperm Receptor Activity (SRA)

In order to evaluate the molecular basis of ZP3 SRA, we subjected the purified glycoprotein to chemical and enzymatic dissection and then tested its components in the "competition assay" [8,13,30–32; J. Bleil and P. Wassarman, unpublished results]. Our observations, summarized below, strongly suggest that ZP3 SRA is attributable to a specific class of serine/threonine-linked (O-linked) oligosaccharides on the glycoprotein. Furthermore, they strongly suggest that galactose, located in α-linkage at the nonreducing terminus of ZP3 O-linked oligosaccharides, ist at least one of the sugar determinants on ZP3 responsible for binding of sperm to the ZP.

Exposure of ZP3 to denaturants, detergents, reducing agents, or elevated temperatures did not significantly alter its SRA, suggesting involvement of oligosaccharides rather than polypeptide chain.

To assess the role of polypeptide chain, ZP3 was digested extensively with insolubilized Pronase, giving rise to glycopeptides (1500–6000 M_r), and these were tested for SRA. The glycopeptides were found to be virtually as effective as intact glycoprotein in inhibiting binding of sperm to eggs (i.e., SRA retained). These results demonstrate that the integrity of ZP3 polypeptide chain is not essential for SRA.

Selective removal of asparagine-linked (N-linked) oligosaccharides from ZP3 by digestion with endo-β-acetyl-D-glucosaminidase F (Endo F) had no effect on its SRA. The O-glycosylated polypeptide chain was found to be virtually as effective as intact glycoprotein (O- and N-glycosylated) in inhibiting binding of sperm to eggs (i.e., SRA retained). These results eliminate the possibility of a role for ZP3 N-linked oligosaccharides in SRA.

Extensive deglycosylation of ZP3 with trifluoromethanesulfonic acid (TFMS), a reagent that removes both N- and O-linked oligosaccharides from glycoproteins, converted ZP3 to a species with an apparent M_r approximating that of the polypeptide chain (44000); this species had no effect on the binding of sperm to eggs (i.e., SRA lost). Taken together with the results described in the previous paragraph this suggests the involvement of ZP3 O-linked oligosaccharides in SRA.

Mild alkaline hydrolysis of ZP3 (β-elimination) selectively removed O-linked oligosaccharides without breakage of peptide bonds (serine and threonine converted to alanine and α-aminobutyric acid, respectively), resulting in an N-glycosylated polypeptide chain; this species had no effect on the binding of sperm to

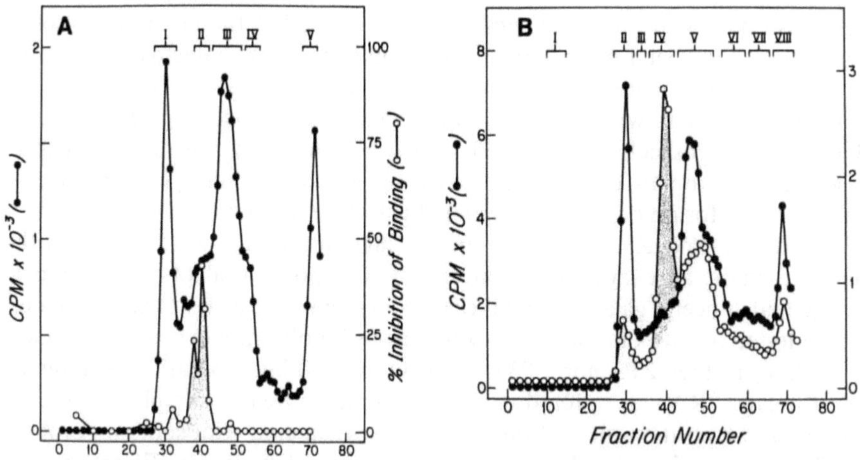

Fig. 13. **A** Gel-filtration and competition binding analyses of O-linked oligosaccharides released from ZP3. Shown are the Bio-Gel P-6 elution profile for radiolableled ([^{3}H]-$NaBH_4$) oligosaccharides released from ZP3 by mild alkaline reduction *(filled circles)* and the SRA profile for the released oligosaccharides *(open circles)*. The region of the elution profile displaying SRA (region II) is *stippled*. **B** Gel-filtration, competition binding analyses, and determination of sperm association of O-linked oligosaccharides released from ZP3. These experiments were carried out as described in **A**. Shown are the Bio-Gel P-6 elution profile for radiolabeled oligosaccharides not incubated with sperm *(filled circles)* and those associated with sperm after a 1 h incubation *(open circles)*. The region of the elution profile that was significantly enriched following incubation of oligosaccharides with sperm (region IV; equivalent to region II in **A**) is *stippled*. Only region IV exhibited sperm receptor activity, reducing sperm binding by more than 50% as compared with controls. (From [31])

eggs (i.e., SRA lost). These results provide direct evidence for the involvement of ZP3 O-linked oligosaccharides in SRA.

Mild alkaline hydrolysis of ZP3 (β-elimination) in the presence of sodium borohydride resulted in release of intact O-linked oligosaccharides having *N*-acetyl-D-galactosaminitol at their reducing termini; these purified oligosaccharides inhibited binding of sperm to eggs (i.e., SRA retained). SRA was only associated with O-linked oligosaccharides having an apparent M_r of 3900, and this size class of oligosaccharides was significantly enriched following affinity purification of alkali-released oligosaccharides with sperm (Fig. 13). These results extend the evidence provided in the previous paragraph, confirming the involvement of ZP3 O-linked oligosaccharides in SRA.

Purified ZP3 and ZP3-derived O-linked oligosaccharides (recovered by β elimination of ZP3 in the presence of sodium borohydride) were treated with a variety of exoglycosidases and the resultant products tested for SRA in the "competition assay." Of six exoglycosidases tested for their ability to destroy SRA, only α-galactosidase and α-fucosidase had a significant inhibitory effect on SRA; β-galactosidase, β-glucuronidase, β-*N*-acetyl-glucosaminidase (containing β-*N*-acetyl-galactosaminidase), and neuraminidase were completely ineffective. Consistent with these observations, much lower levels of sperm binding were observed with formaldehyde-fixed, ovulated eggs that had been exposed to either α-galactosidase or α-fucosidase than with untreated eggs and eggs exposed to the other exoglycosidases. These results suggest that removal of either terminal galactose or fucose from ZP3 O-linked oligosaccharides destroys glycoprotein SRA. However, it is important to note that, while we were able to detect the release of galactose from ZP3 O-linked oligosaccharides treated with α-galactosidase, we did not detect release of fucose from α-fucosidase treated samples. Consequently, we can make a strong case only for involvement of terminal galactose in SRA.

Purified ZP3 and ZP3-derived O-linked oligosaccharides were treated with galactose oxidase, an enzyme that converts the C-6 position alcohol of terminal galactose and *N*-acetyl-galactosamine residues in oligosaccharides to an aldehyde, and then tested for SRA in the "competition assay." Following exposure to galactose oxidase, neither ZP3 nor ZP3-derived O-linked oligosaccharides exhibited SRA. However, SRA was restored to galactose oxidase treated samples by regeneration of a C-6 position alcohol on galactose by reduction with sodium borohydride. Thus, conversion of a ZP3 sugar alcohol to an aldehyde is sufficient to preclude binding of sperm to ZP3. These results are not unexpected, since interactions between carbohydrate-binding proteins and sugar ligands are stabilized largely by hydrogen bonds and van der Waals contacts, with the former providing the major contribution to binding. Sugar hydroxyls participate extensively in hydrogen bonding, probably accounting for the inhibitory effects of galactose oxidase treatment on carbohydrate-mediated sperm-egg interaction.

Finally, it is tempting to speculate that in mice, as in sea urchins [33,34], a lectin-like protein associated with sperm ("egg-binding protein") binds to sperm receptor oligosaccharides, with terminal galactose serving as at least one of the binding determinants. In this connection, it is interesting to note that oligosaccharides with repeating *N*-acetyl-lactosamine units (i.e., lactosaminoglycans) are usually associated with developmentally regulated glycoconjugates [35,36]. In a number of cases (e.g., neurons and erythrocytes), such sequences are often terminated by galactose in α linkage with the galactose of the *N*-acetyl-lactosamine unit [37,38]. Inactivation of ZP3 as a sperm receptor, which occurs following fertilization, could be due to removal or modification of one or more sugars of ZP3 O-linked oligosaccharides involved in sperm binding. Our results demonstrate that removal or modification of even a single sugar that serves as a sperm binding determinant, leads to loss of SRA. It is likely that cortical granule enzyme(s) is responsible for inactivation of ZP3 following fertilization [1-5].

Role of Polypeptide Chain in Acrosome Reaction Inducing Activity

Whereas the ability of ZP3 to serve as a sperm receptor depends solely on its O-linked oligosaccharides (discussed above), the ability of ZP3 to serve as an AR inducer depends on its polypeptide chain as well [8,30]. We found that, although small (1500-6000 M_r) glycopeptides and O-linked oligosaccharides derived from ZP3 were able to bind to heads of acrosome-intact sperm and prevent them from binding to ovulated eggs, neither the glycopeptides nor O-linked oligosaccharides induced sperm to undergo the AR in vitro. However, large proteolytic fragments of ZP3 ($\simeq$40000 M_r) were able to induce the AR in vitro. In view of this behavior, we suggest that sperm bind to multiple O-linked oligosaccharides on each ZP3 molecule (i.e., a multivalent interaction), and that these oligosaccharides must be joined by polypeptide chain for induction of the AR. The distance between these oligosaccharides on an individual ZP3 molecule may be an important factor during induction of the AR. It should be noted, however, that we have not excluded the possibility that a portion of the polypeptide chain itself functions as a "fusion peptide."

Summary and Final Comments

Here, we have reviewed research on the mouse sperm receptor carried out in our own laboratory. Among major conclusions to be drawn from this research are the following:

1. The mouse sperm receptor is an approximately 83000 M_r glycoprotein, called ZP3, that is located in egg ZP in more than a billion copies. It is one of three glycoproteins (ZP1, ZP2, and ZP3) that make up mouse ZP.
2. ZP3 is synthesized (and ZP3 gene expressed) and secreted exclusively by oocytes during their growth phase. ZP3 synthesis (and ZP3 gene expression) is terminated at the time of ovulation.
3. ZP3 consists of a approximately 44000 M_r polypeptide chain (402 amino acids; mature ZP3), three or four complex-type N-linked oligosaccharides, and an undetermined number of O-linked oligosaccharides. ZP3 oligosaccharides are responsible for the molecular heterogeneity of the glycoprotein and, in large part, for its acidic nature.
4. ZP3 is part of a heterodimer (ZP2: ZP3; 180000 M_r), held together by noncovalent interactions, that conitutes the repeating (every 14 nm or so) structural unit (monomer) of ZP filaments (polymer). Therefore, in addition to its roles during fertilization, ZP3 plays a structural role in ZP assembly.
5. ZP3 serves as inducer of the AR following binding of sperm to ZP.
6. ZP3 is inactivated following fertilization so that it no longer exhibits either SRA- or AR-inducing activity in vitro. Inactivation of ZP3 is probably attributable to cortical granule contents that are deposited into the perivitelline space following fertilization.
7. A specific size class of ZP3 O-linked oligosaccharides ($\simeq 3900$ M_r) is responsible for glycoprotein SRA. Galactose, located in α linkage at the nonreducing terminus of ZP3 O-linked oligosaccharides, is at least one of the determinants responsible for binding of sperm to ZP3.
8. The ZP3 polypeptide chain plays a role in glycoprotein AR-inducing activity, but not in its SRA.

The mouse sperm receptor ZP3 illustrates a role for carbohydrates in supporting specific interactions between eukaryotic cells. A strong case can be made for involvement of carbohydrates in induction of the AR and species-specific binding of sperm to eggs in marine invertebrates as well. Apparently, only a relatively small percentage of ZP3 O-linked oligosaccharides are actually recongnized by acrosome-intact sperm, but each sperm probably binds to multiple O-linked oli gosaccharides on each ZP3 molecule. It is possible that induction of the AR i dependent on multivalent interactions between egg-binding proteins on the sperr head and ZP3. Whether ZP3 or O-linked oligosaccharides derived from ZP exhibit any species specificity has not yet been determined. The great diversity o known oligosaccharide structures is certainly compatible with generation of spe cies specificity for mammalian sperm receptors. The variety of composition sequences, branching patterns, conformations, and other features of oligosaccha ride structure provide for a staggering number of possible combinations. It will b of considerable interest to compare the structures of these oligosaccharides wit functionally analogous oligosaccharides isolated from eggs of other mammals.

References

1. Gwatkin RBL (1977) Fertilization mechanisms in man and mammals, Plenum, New York
2. Yanagimachi R (1981) In: Mastroianni L, Biggers JD (eds) Fertilization and embryonix deve opment In vitro, Plenum, New York, p 81
3. Yanagimachi R (1988) In: Knobil E, Neill JD (eds) The physiology of reproduction, Rave New York, vol 1, p 135
4. Wassarman PM (1987) Science 235: 553
5. Wassarman PM (1987) Annu Rev Cell Biol 3: 109
6. Wassarman PM (1983) In: Yamada K (ed) Cellular interactions and development: molecul mechanisms, Wiley, New York, p 1
7. Wassarman PM, Florman HM, Greve JM (1985) In: Metz CB, Monroy A (eds) Biology of fe tilization, Academic, New York, vol 2, p 341
8. Wassarman PM, Bleil JD, Florman HM, Greve JM, Roller RJ, Salzmann GS (1985) Co Spring Harbor Symp Quant Biol 50: 11
9. Wassarman PM, Bleil JD, Florman HM, Greve JM, Roller RJ, Salzmann GS (1986) In: Gal (ed) Gametogenesis and the early embryo, Liss, New York, p 371
10. Wassarman PM, Bleil JD, Florman HM, Greve JM, Roller RJ, Salzmann GS (1986) I Hedrick JL (ed) The molecular and cellular biology of fertilization, Plenum, New York, p 55
11. Wassarman PM, Bleil JD, Fimiani CM, Florman HM, Greve JM, Roller RJ Salzmann G Samuels FG, Wenk-Salamone K (1987) In: Firtel RA, Davidson EH (eds) Molecul approaches to developmental biology, Liss, New York, p 51
12. Bleil JD, Wassarman PM (1980) Dev Biol 76: 185
13. Wassarman PM (1988) Annu Rev Biochem 57: 415
14. Bleil JD, Wassarman PM (1980) Cell 20: 873
15. Saling PM, Sowinski J, Storey BT (1979) J Exp Zool 209: 229
16. Florman HM, Storey BT (1979) Dev Biol 91: 121
17. Bleil JD, Wassarman PM (1983) Dev Biol 95: 317
18. Bleil JD, Wassarman PM (1986) J Cell Biol 102: 1363
19. Vazquez MH, Wassarman PM (1987) J Cell Biol 105: 5a
20. Greve JM, Wassarman PM (1985) J Mol Biol 181: 253
21. Bleil JD, Wassarman PM (1980) Proc Natl Acad Sci, USA 77: 1029
22. Greve JM, Salzmann GS, Roller RJ, Wassarman PM (1982) Cell 31: 749
23. Salzmann GS, Greve JM, Roller RJ, Wassarman PM (1983) EMBO J 2: 1451
24. Roller RJ, Wassarman PM (1983) J Biol Chem 258: 13243
25. Ringuette MJ, Sobieski DA, Chamow SM, Dean J (1986) Proc Natl Acad Sci, USA 83: 4341
26. Kinloch RA, Roller RJ, Fimiani CM, Wassarman DA, Wassarman PM (1988) Proc Natl Ac Sci, USA (in press)
27. von Heijne G (1985) J Mol Biol 184: 99
28. Andrews PC, Nichols R, Dixon JE (1987) J Biol Chem 262: 12692

29. Struck DK, Lennarz WJ (1980) In: Lennarz WJ (ed) The biochemistry of glycoproteins and proteoglycans, Plenum, New York, p 35
30. Florman HM, Bechtol KB, Wassarman PM (1984) Dev Biol 106: 243
31. Florman HM, Wassarman PM (1985) Cell 41: 313
32. Bleil JD, Wassarman PM (1988) Proc Natl Acad Sci, USA (in press)
33. Lopo AC, Glabe CG, Lennarz WJ, Vacquier VD (1982) Ann NY Acad Sci 383: 405
34. Ruiz-Bravo N, Rossignol DP, Decker GL, Rosenber LI, Lennarz WJ (1986) In: Hedrick JL (ed) The molecular and cellular biology of fertilization, Plenum, New York, p 293
35. Pink JRL (1980) Contemp Topics Mol Immunol 9: 89
36. Feizi T, Kapadia A, Gooi HC, Evans MJ (1981) In: Muramatsu T, Ikawa Y (eds) Teratocarcinoma and cell surface North Holland, Amsterdam, p 167
37. Dodd J, Jessell TM (1985) J Neurosci 5: 3278
38. Jessell TM, Dodd J (1985) Phil Trans Roy Soc Lond (Biol) 308: 271

Ovarian Development and the Formation of the Mammalian Zona Pellucida*

B. S. Dunbar, G. A. Maresh, and K. Washenik

Introduction

The mammalian zona pellucida (ZP) is the unique extracellular matrix which i formed during oocyte growth and development. It is well established that thi matrix is important in the fertilization process because the sperm must frist adher to the surface of the ZP and penetrate it before it can reach the membrane of th ovum (see reviews by Yanagimachi 1977; Dunbar 1983a; Dunbar and Wolgemut 1984; Wassarman 1987). This structure subsequently aids in the maintenance c cell contact of the cleaving blastomeres and also protects the embryo during it transit down the oviduct to the uterus (Mintz 1962). Additional studies demor strate that antibodies to the ZP will inhibit or reduce fertilization following eithe passive or active immunization (East et al. 1984, 1985; Sacco et al. 1981; Wood e al. 1981). There have therefore been many investigations which have led to th detailed characterization of the constituent macromolecules of the ZP. In additio these studies have led to the observations that there are marked variations in th morphological, biochemical, immunochemical, and physicochemical properties c the constituent glycoproteins of the ZP in different mammalian species (Bleil an Wassarman 1980a-c; Dunbar et al. 1981, 1985; Sacco et al. 1981; Drell and Dur bar 1984; Timmons et al. 1987; Yurewicz et al. 1987; Maresh and Dunbar 1987a One major morphological variation is in the thickness of the ZP in different mar malian species, as illustrated in Fig. 1 (see also review by Dunbar 1983a). In add tion to their size, the differences in sensitivity to enzymatic digestion as dete mined by visual dissolution using microscopy is also impressive. These difference may reflect the variation in the relative amount of protein per zona pellucida (e.g 4.8 ng per mouse ZP. Bleil and Wassarman 1980a, b; 30 ng per pig ZP, Dunbar al. 1981). The protein patterns of ZP resolved by one- and two-dimensional pol acrylamide gel electrophoresis (PAGE) also vary (see review by Dunbar 1983b Timmons and Dunbar 1987). The comparison of the glycosylated and deglycos lated ZP proteins of pig and rabbit ZP are illustrated in Fig. 2 and 3. Finally, var ations in the composition of ZP in different species (Dunbar et al. 1981; Sacco al. 1981) appear to be the reason for the complex properties of the antigenicity a well as immunogenicity of ZP antigens in different species (Sacco et al. 198 Drell and Dunbar 1984; Maresh and Dunbar 1987a; see also reviews by Dunb

* Work funded by grants to B. S. Dunbar from the Mellon Foundation and the National Institu of Child Health and Human Development.

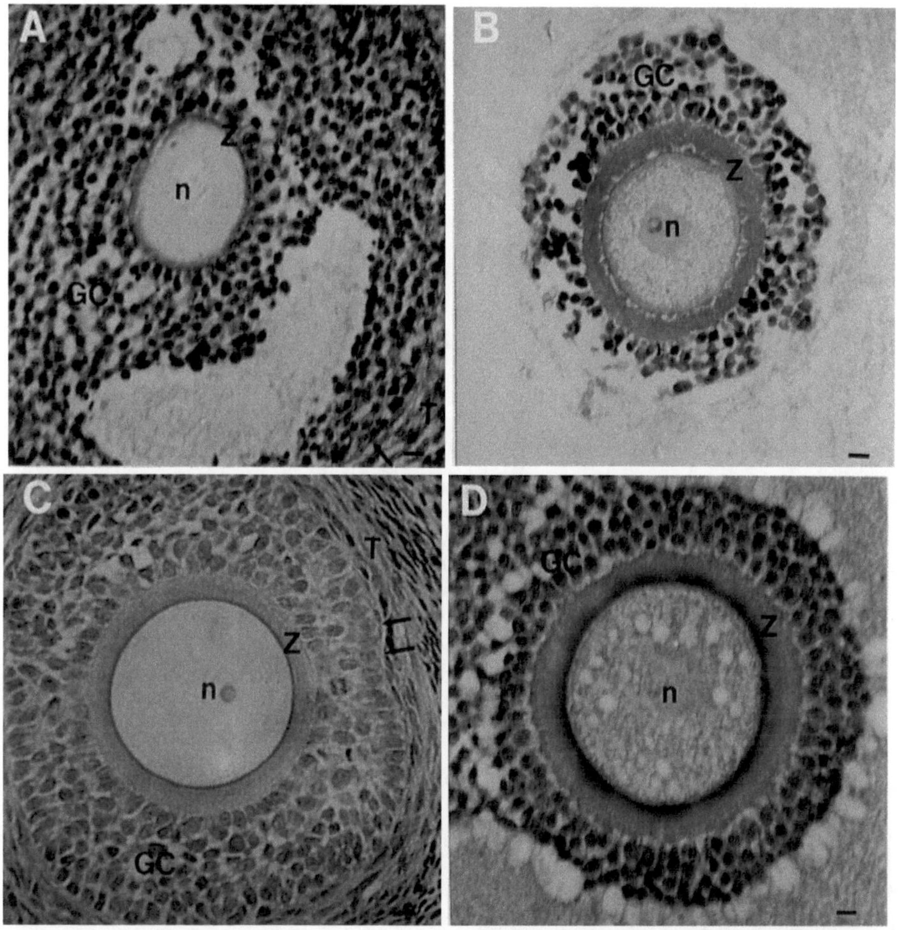

Fig. 1 A–D. Light micrographs of oocytes in tertiary or antral follicles. **A** Rat. **B** Dog. **C** Baboon. **D** Cat. Note the comparison of the size of the zona pellucida in follicles of different mammalian species and the distinct layers in the baboon and cat ZP. *Z*, Zona pellucida; *GC*, granulosa cells; *arrows*, basal lamina; *T*, thecal cells; *n*, nucleus; *bars* 15 μm

et al. 1986; Timmons and Dunbar 1987; and contributions by other authors in this text).

The biochemical composition of some species of mammalian ZP has been studied extensively, and in the mouse the synthesis of the ZP has been studied as well (Bleil and Wassarman 1980c; Shimizu et al. 1983; Philpott et al. 1987). However, investigations on the extracellular assembly of the constituent ZP molecules and their potential roles in early follicular development have been limited. In order to design studies to investigate the potential roles of the ZP during follicular development it is necessary to develop a basic understanding of some of the events which occur during ovarian cell differentiation.

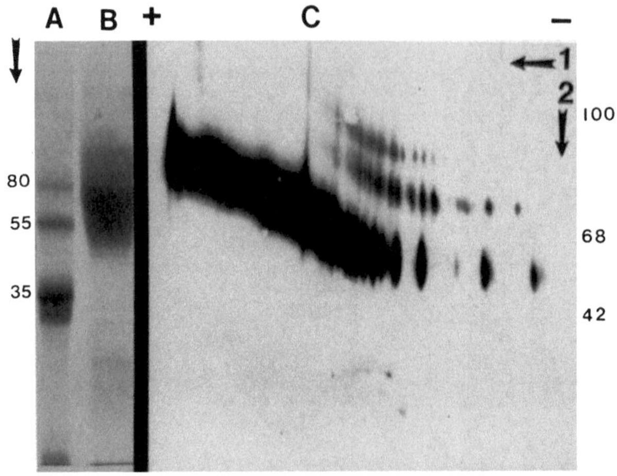

Fig. 2. Pig ZP glycoproteins chemically deglycosylated with trifluoromethanesulfonic acid *(A* compared with nondeglycosylated glycoproteins *(B)*, analyzed by one-dimensional PAGE and sil· ver staining ($M_r = \times 10^{-3}$). Two-dimensional PAGE of glycosylated pig ZP glycoproteins *(C)* Note extreme molecular weight and charge heterogeneity of ZP protein pattern *(B* and *C)*. The heterogeneity is reduced if proteins are deglycosylated as shown in lane *A*

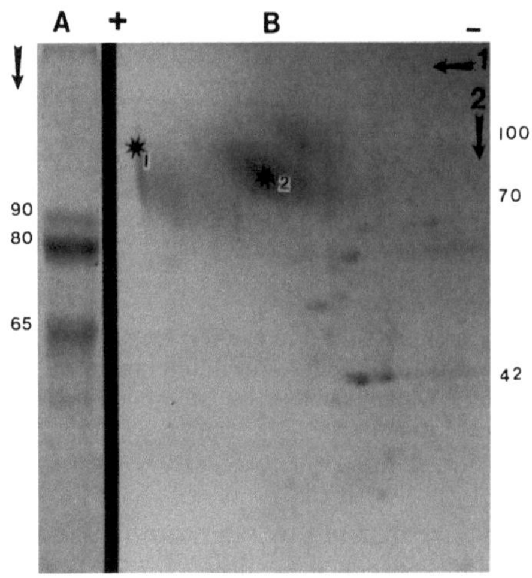

Fig. 3. Rabbit ZP glycoproteins chemically deglycosylated with trifluoromethane sulfonic acid analyzed by one-dimensional PAGE and silver staining ($M_r = \times 10^{-3}$) *(A)*. Two-dimensional PAGE of glycosylated rabbit ZP glycoproteins *(B)*; note extreme molecular weight and charge heterogeneity of ZP protein pattern. The heterogeneity is reduced if proteins are deglycosylated as shown in lane *A*. *Asterisks*, areas of gels excised and used to produce antibodies to rabbit ZP glycoproteins 1 and 2

Stages of Follicular Development

The mature preovulatory ovarian follicle consists of the oocyte surrounded by (a) the zona pellucida, (b) multiple layers of granulosa cells, and (c) a basal lamina which separates the granulosa cells from the theca interna and externa (see Fig. 1). To date, many investigations have been carried out to study the regulation and differentiation of follicular cells during the stage of antrum formation (see reviews by Hillier 1985; Richards 1980). However, studies on earlier stages of follicular development during which the ZP is formed in either the sexually mature or immature animal are limited.

During early differentiation of the fetal ovary many of the somatic cells which will form the mature follicle develop from the mesonephric tissue (Byskov et al. 1985; Byskov 1986). Divergent patterns of ovarian development are apparent in different animal species, including the differentiation of various steroid-producing cells (Byskov et al. 1985) as well as oogenesis (see reviews by Lamming 1984; Austin and Short 1982; Byskov 1986).

Oogenesis, which includes the formation, growth, and maturation of the female gamete, is initiated in embryonic life and continues at dissimilar rates and stages in different animal species (Mauleon and Mariana 1977; Lavasseur 1979), as does folliculogenesis, which is the process of somatic cells becoming associated with the oocyte to form a follicular unit. In some animal species (e.g., cow, sheep, and human) oogenesis and folliculogenesis begin before birth, in some (rabbit and hamster) these processes begin following birth, while in others (rat, mouse, and pig) oogenesis begins before birth. Folliculogenesis, however, is not initiated until after birth. Furthermore, the time from the initiation of oogenesis to the onset of folliculogenesis can vary from as little as 5-7 days (mouse, rat) to over 50 days in other species (human, cow, pig). To date, little is known about the regulation of the onset of oogenesis and folliculogenesis or about early events that accompany the growth of the oocyte and surrounding follicle cells which are destined to differentiate into the steroid-producing granulosa cells. Because the molecules of the ZP are produced and secreted during these early stages of oocyte growth and follicular development, they provide excellent markers for this dynamic process (Oakberg and Tyrell 1975; Wolgemuth et al. 1984; Philpott et al. 1987). In studies by Oakberg and Tyrell (1975) in which the ZP has been used as a marker, it has been estimated that the development of a follicle extends over a 5-week period in mice. While this information is not yet available for other species, it is likely that the time during which the entire zona matrix is formed spans several days to weeks. While a considerable amount of information is now available on the structure and function of the mature zona pellucida obtained from large ovarian follicles or from ovulated ova with respect to their potential roles in the fertilization process, few studies have been carried out to evaluate the potential roles of the zona pellucida proteins during ovarian follicular development or during the onset of sexual maturation. The initial studies into the molecular nature of the ZP are, however, prerequisite to our understanding of these developmental processes.

Because large quantities of ZP can be easily isolated from both pig and rabbit ovaries, the analysis of the constituent glycoproteins and the preparation of specific polyclonal and monoclonal antibodies to these molecules has been possible

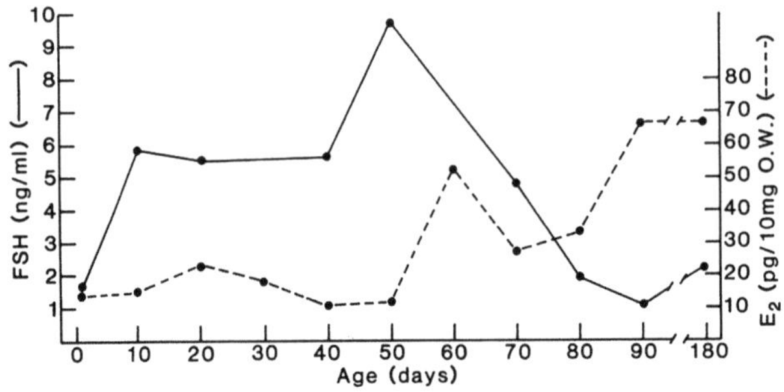

Fig. 4. Summary of hormonal profile in rabbit during different stages of sexual maturation. *FSH,* follicle stimulating hormone analyzed in the serum; E_2, estradiol measured in the ovaries (pg/10 mg ovarian weight)

(Timmons et al. 1987; Drell and Dunbar 1984; Koyama et al. 1985; Mori et al. 1985). Furthermore, the rabbit has proven to be an excellent model system for studies of the early events of ovarian follicular growth since it has a prolonged stage of follicular development following onset of oogenesis which occurs after birth. Because of this prolonged period postpartum it is easy to obtain ovaries with follicles at more defined stages of development.

A great deal of information concerning the hormonal status of these animals is already available (de Turckheim et al. 1983) which can be related to stages of follicular development and the formation of the zona pellucida. The hormonal status of rabbits of different ages is summarized in Fig. 4. A histological section of a rabbit ovary obtained at approximately 6 weeks (Fig. 5A) shows abundant primordial and primary follicles. It is also interesting to note that at this stage rabbits have an elevated endogenous follicle-stimulating hormone (FSH) level: the administration of exogenous gonadotrophins to these young animals will not induce the recruitment of any of these follicles to the antral stage (Hertz 1986; Tavella and Dunbar, unpublished observations). However, at 12 weeks, when the serum FSH levels are lower, numerous secondary and antral follicles are present (Fig. 5B), can be recruited, and will be ovulated when exogenous FSH and human chorionic gonadotropin (HCG) are administered. Even though numerous antral follicles are present at this stage, spontaneous ovulation will not occur unless exogenous hormones are given until the rabbit is sexually mature at the age of 4–6 months (de Turckheim et al. 1983; Kamwanja and Hauser 1983). At this stage of ovarian development which accompanies the onset of sexual maturation, the ovary has developed a predominant interstitial cell gland which is also a steroid-producing organ (see Fig. 5C).

Fig. 5 A-C. Light micrograph of an ovary from a 6-week-old (**A**), 12-week-old (**B**), and sexually mature rabbit (**C**). **PF,** primordial follicles; *1°*, primary follicle; *2°*, secondary follicle; *A,* antral follicle; *CL,* corpus luteum; *IC,* interstitial cells

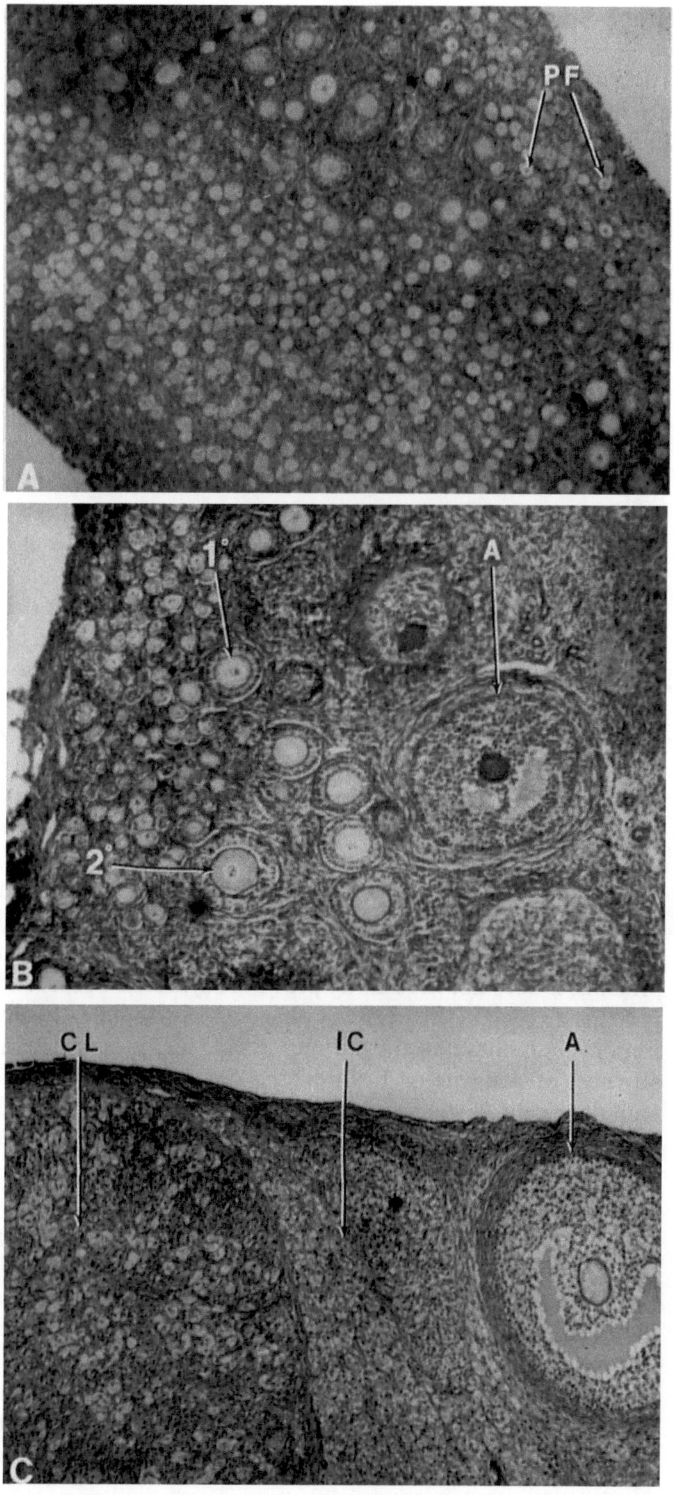
PF
A
1°
A
2°
B
CL
IC
A
C

Stage-Specific Development and the Formation of the Zona Pellucida

The synthesis of the constituent molecules of the zona pellucida and their assembly into an extracellular matrix have been studied using a variety of methods. Early studies relied predominantly on microscopic analysis to examine the extracellular form of this matrix (Kang 1974). These studies demonstrated that the diameter of the ZP is enlarged as the oocyte grows, and the surrounding follicle cells increase in number and differentiate into steroid secreting cells. The inner layers continue to maintain contact with the maturing oocyte through junctional complexes (Anderson and Albertini 1976). Studies further demonstrate that the ZP of some species appear to contain multiple layers (Konecny 1959; Hope 1965; Wolgemuth et al. 1984; see also Fig. 1).

To date, it is evident that the mouse oocyte is capable of synthesizing ZP glycoproteins in vitro (Bleil and Wassarman 1980c; Shimizu et al. 1983). More recently, in situ hybridization studies have also been used to detect RNA coding for the mouse ZP protein, which has been identified as the sperm receptor. Immunocytochemical localization methods have also been used to identify ZP antigens in the cytoplasm of hamster oocytes (Leveille et al. 1987) and rabbit oocytes (Wolgemuth et al. 1984).

While it is well accepted that the oocyte is involved in the synthesis of ZP glycoproteins, the role of the granulosa cells in the formation of the ZP remains less clear. Electron micrographs of the cytoplasm of granulosa cells adjacent to the oocyte in primary follicles show pockets of material similar in appearance to the newly forming extracellular material which will become the ZP (see Baker 1971). Recently additional evidence for the synthesis of ZP glycoproteins by granulosa cells has been presented. Immunocytochemical localization of ZP glycoproteins in the granulosa cells has been demonstrated in the rabbit (Wolgemuth et al. 1984; see Fig. 7; Skinner and Dunbar, unpublished observations) as well as the pig (Hedrick and Fry 1980) and hamster (Leveille et al. 1987). These studies have been extended to demonstrate that different rabbit ZP antigens are localized at different stages of follicular development (Fig. 6). Although localization implies synthesis of proteins, it does not necessarily prove the synthesis because cells may also take up proteins by such mechanisms as phagocytosis. To date, the majority of these localization studies have used light-microscopic methods to determine the presence (or absence) of antigens or RNA in the cell cytoplasm. Because the cytoplasm of granulosa cells is sparse when compared to oocyte cytoplasm, light-microscopic methods are frequently inadequate. The use of electron-microscopic methods in conjunction with specific probes will be valuable for studying these processes in greater detail.

More recently we have been able to establish that long-term cultures (>2 months) of granulosa cells cultured in vitro from primary and early secondary follicles secrete ZP antigens into the media (Maresh and Dunbar 1987b). It is further observed that two different ZP antigens are expressed if cultured on different extracellular matrices.

Indirect evidence that granulosa cells produce the ZP matrix also exists. When rabbits are immunized with ZP proteins, the oocytes are destroyed at the stage at which they begin to initiate formation of the zona pellucida matrix (see discussion

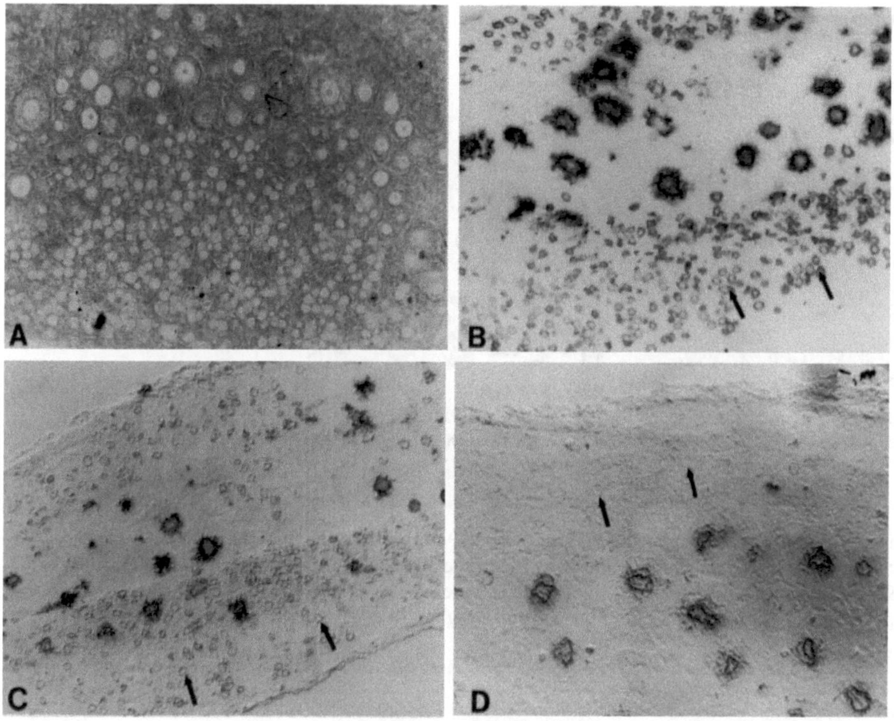

Fig. 6 A-D. Immunocytochemical localization in frozen sections of 6-week-old rabbit ovaries of ZP antigens. At this age, most of the follicles are growing, although they will eventually undergo atresia. **A** Control section stained with hemotoxalin and eosin. **B** Sheep anti-heat-solubilized rabbit ZP. **C** Sheep anti-80-kd ZP protein (see asterisk "1", Fig. 3). **D** Sheep anti-65-kd ZP protein (see asterisk "2", Fig. 3). Frozen sections analyzed with modulation contrast optics. Note staining in growing primordial follicles *(arrows)* in **B** and **C**, but not in **D**. The antibody which recognizes the deglycosylated 80-kd ZP protein stains only the later stage (growing) follicles *(D)*

in Skinner et al. 1984). Even though there are no remnants of oocytes, there are distinct matrices which stain identical to ZP surrounded by "granulosa" cells (see Fig. 8). These matrices are easily distinguished from the collapsed remnants of ZP from atretic follicles which do not contain granulosa cells within the center of the collapsed zona pellucida. It therefore appears from these studies that at least some ZP proteins are produced in the absence of the oocytes in vivo. It remains to be determined whether the apparent contribution of the granulosa cells to the formation of the ZP occurs only in mammalian species in which this matrix is thicker and contains multiple layers.

In summary, it is clear that the formation of the zona pellucida matrix is a complex process which occurs during different stages of ovarian follicular development. It is evident that this process is dependent on the species, which reflects the variations in the properties of different ZP molecules.Given the major advances in protein isolation and characterization of the zona pellucida molecules and the continued development of specific antibodies and cDNA probes to the ZP

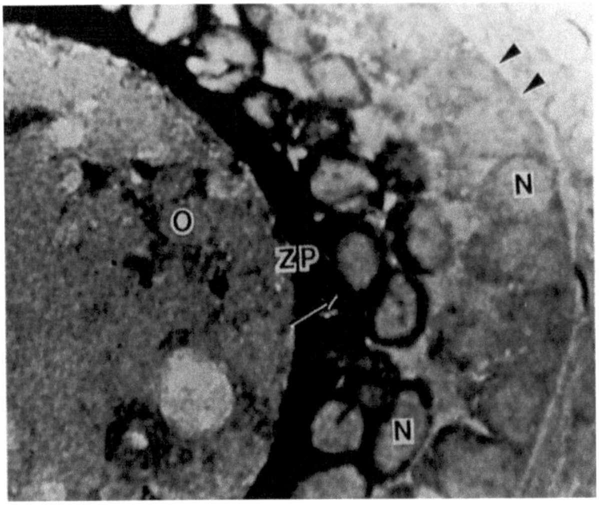

Fig. 7. Immunocytochemical localization of ZP antigens in the cytoplasm of the inner layer of rabbit granulosa cells using antibody to two-dimensional PAGE-purified rabbit ZP protein. Over 96% of follicles at this stage exhibit staining in these cells *(arrow)*. *O*, oocyte; *N*, nucleus; *ZP* zona pellucida; *arrowheads*, basal lamina. (From Wolgemuth et al. 1984)

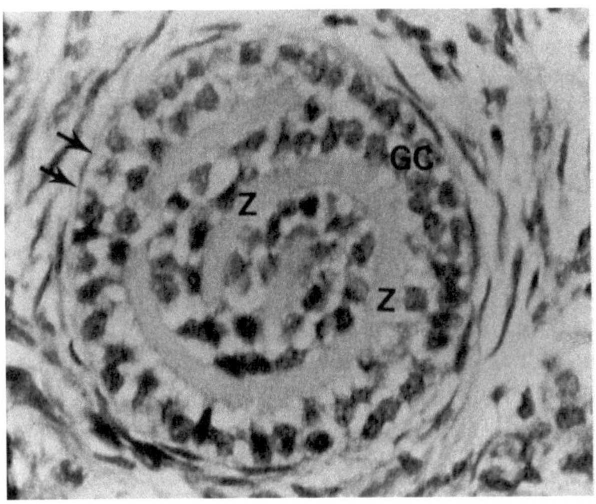

Fig. 8. Light micrograph of an ovary from a rabbit immunized with pig ZP proteins in which oocytes are destroyed. "Zona pellucida-like" matrix is observed within a spiral of granulosa cells. This type of structure is abundant in follicles during the early stages of immunization. *Arrows*, basal lamina; *Z*, structure having identical staining properties to zona pellucida; *GC*, granulosa cells

of different species, it is likely that a better understanding of this dynamic process will soon be forthcoming.

Acknowledgements. We wish to thank Ms. Suzanne Mascola for her expertise in typing the manuscript and Ms. Debbie Delmore for assistance in photography.

References

Anderson E, Albertini DF (1976) Gap junctions between the oocyte and companion follicle cells in the mammalian ovary. J Cell Biol 71: 680-686

Austin CR, Short R (eds) (1982) Reproduction in mammals, vol 1, 2nd edn Cambridge University Press, Cambridge

Baker TG (1971) Electron microscopy of the primary and secondary oocyte. In: Raspe G (ed) Advances in biosciences, vol 6. Vieweg, Brunswick; Pergamon, New York, pp 6-23

Bleil JD, Wassarman PM (1980a) Mammalian sperm-egg interaction: identification of a glycoprotein in mouse egg zonae pellucidae possessing receptor activity for sperm. Cell 20: 873-882

Bleil JD, Wassarman PM (1980b) Structure and function of the zona pellucida: identification and characterization of the proteins of the mouse oocyte's zona pellucida. Dev Biol 76: 185-202

Bleil JD, Wassarman PM (1980c) Synthesis of zona pellucida proteins by denuded and follicle-enclosed mouse oocytes during culture in vitro. Proc Natl Acad Sci USA 77: 1029-1033

Byskov AG (1986) Differentiation of the mammalian embryonic gonad. Physiol Rev 66 (1): 71-117

Byskov AG, Hoyer PE, Westergaard L (1985) Origin and differentiation of the endocrine cells of the ovary. J Reprod Fertil 75: 299-306

De Turckheim M, Berger M, Jean-Faucher C, Veyssiere G, Jean C (1983) Changes in ovarian oestrogens and in plasma gonadotrophins in female rabbits from birth to adulthood. Acta Endocrinol (Copenh) 103: 125-130

Drell DW, Dunbar BS (1984) Monoclonal antibodies to rabbit and pig zonae pellucidae distinguish species-specific and shared antigenic determinants. Biol Reprod 30: 445-457

Dunbar BS (1983a) Antibodies to zona pellucida antigens and their role in fertility. In: Wegman TG, Gill TJ III (eds) Immunology of reproduction. Oxford University Press, London, pp 505-534

Dunbar BS (1983b) Morphological, biochemical and immunochemical characterization of the mammalian zona pellucida. In: Hartmann J (ed) Mechanism and control of animal fertilization. Academic, New York, pp 140-175

Dunbar BS, Wolgemuth D (1984) Structure and function of the mammalian zonae pellucidae. In: Satir B (ed) Modern cell biology, vol 3. Liss, New York, pp 77-111

Dunbar BS, Liu C, Sammons DW (1981) Identification of the three major proteins of porcine and rabbit zonae pellucidae by two-dimensional gel electrophoresis: comparison with follicular fluid, sera and ovarian cell proteins. Biol Reprod 24: 1111-1124

Dunbar BS, Dudkiewicz AB, Bundman DS (1985) Proteolysis of specific porcine zona pellucida glycoproteins by boar acrosin. Biol Reprod 32: 619-630

Dunbar BS, Timmons TM, Maresh GA, Niu EM, Skinner SM (1986) In: Clark DA, Croy BA (eds) Reproductive immunology 1986. Elsevier, Amsterdam, pp 116-124

East IJ, Mattison DR, Dean J (1984) Monoclonal antibodies to the major protein of the murine zona pellucida: effects on fertilization and early development. Dev Biol 104: 49-56

East IJ, Guylas BJ, Dean J (1985) Monoclonal antibodies to the murine zona pellucida protein with sperm receptor activity: effects on fertilization and early development. Dev Biol 109: 268-273

Hedrick JL, Fry GN (1980) Immunocytochemical studies on the porcine zona pellucida. J Cell Biol 87: 136a

Hertz R (1986) Whence the antihormones? In: Talwar GP (ed) Immunological approaches to contraception and promotion of fertility. Plenum, New York, pp 5-15

Hillier SG (1985) Sex steroid metabolism and follicular development in the ovary. In: Clarke JR (ed) Oxford review of reproductive biology. Oxford University Press, New York, pp 168-222

Hope J (1965) The fine structure of developing follicles of the rhesus monkey. J Ultrastruct Res 12: 592-610
Kamwanja LA, Hauser ER (1983) The influence of photoperiod on the onset of puberty in the female rabbit. J Anim Sci 56: 1370-1375
Kang Y (1974) Development of the zona pellucida in the rat oocyte. Am J Anat 139: 535-566
Konecny M (1959) Etude histochemique de la zone pellucide des ovules de chatte. CR Soc Biol (Paris) 153: 893-894
Koyama K, Hasegawa A, Tsuji Y, Isojima S (1985) Production and characterization of monoclonal antibodies to crossreactive antigens of human and porcine zonae pellucidae. J Reprod Immunol 7: 187-198
Lamming GE (1984) Marshall's physiology of reproduction, vol 1, 4th edn, Livingstone, Edinburgh
Lavasseur MC (1979) Thoughts on puberty. The gonads. Ann Biol Anim Biochem Biophys 19: 321-335
Leveille MC, Roberts KD, Chevalier S, Chapdelaine A, Bleau G (1987) Formation of the hamster zona pellucida in relation to ovarian differentiation and follicular growth. J Reprod Fertil 79: 173-183
Maresh GA, Dunbar BS (1987a) Antigenic comparison of five species of mammalian zonae pellucidae. J Exp Zool 244: 299-307
Maresh GA, Dunbar BS (1987b) Expression of specific proteins by ovarian primary follicles cultured in vitro. Biol Reprod [Suppl 1] 36: 88a
Mauleon P, Mariana JC (1977) Oogenesis and folliculogenesis. In: Cole HH, Cupps PT (eds) Reproduction in domestic animals. Academic, New York, pp 175-201
Mintz E (1962) Experimental study of the developing mammalian egg: removal of the zona pellucida. Science 138: 594-597
Mori T, Kamada M, Yamana S, Kinoshita T, Kano K, Mori T (1985) Production of monoclonal antibodies to porcine zona pellucida and their inhibition of sperm penetration through human zona pellucida in vitro. J Reprod Immunol 8: 1-11
Oakberg EF, Tyrell PD (1975) Labeling the zona pellucida of the mouse oocyte. Biol Reprod 12: 477-482
Philpott CC, Ringuette MJ, Dean J (1987) Oocyte specific expression and developmental regulation of ZP3, the sperm receptor of the mouse zona pellucida. Dev Biol 121: 568-575
Richards JS (1980) Maturation of ovarian follicles: actions and interactions of pituitary and ovarian hormones on follicular differentiation. Physiol Rev 60: 51-89
Sacco AG, Yurewicz EC, Subramanian MG, DeMayo FJ (1981) Zona pellucida composition: species cross-reactivity and contraceptive potential of antiserum of a partially purified pig zona antigen. Biol Reprod 25: 997-1008
Shimizu S, Tsuji M, Dean J (1983) In vitro biosynthesis of three sulfated glycoproteins of murine zonae pellucidae by oocytes grown in follicle culture. J Biol Chem 258: 5858-5863
Skinner SM, Mills T, Kirchick HJ, Dunbar BS (1984) Immunization with zona pellucida proteins results in abnormal ovarian follicular differentiation and inhibition of gonadotropin-induced steroid secretion. Endocrinology 115(6): 2418-2432
Timmons TM, Dunbar BS (1987) Antigens of mammalian zona pellucida. In: Matheur S, Fredericks CM (eds) Perspectives of immunoreproduction: Conception and contraception. Hemisphere, Washington (in press)
Timmons TM, Maresh GA, Bundman DS, Dunbar BS (1987) Use of specific monoclonal and polyclonal antibodies to define distinct antigens of the porcine zonae pellucidae. Biol Reprod 36: 1275-1287
Wassarman P (1987) The biology and chemistry of fertilization. Science 5: 2924-2935
Wolgemuth D, Celenza J, Bundman D, Dunbar BS (1984) Formation of the rabbit zona pellucida and its relationship to ovarian follicular development. Dev Biol 106: 1-14
Wood DM, Liu C, Dunbar BS (1981) The effects of alloimmunization and heteroimmunization with zonae pellucidae on fertility in rabbits. Biol Reprod 25: 439-450
Yanagimachi R (1977) Specificity of sperm-egg interaction. In: Edidin M, Johnson MH (eds) Immunobiology of gametes. Cambridge University Press, Cambridge, pp 187-207
Yurewicz EC, Sacco AG, Subramanian MG (1987) Structural characterization of the $M_r = 55{,}000$ antigen (ZP3) of porcine oocyte zona pellucida. J Biol Chem 262(2): 564-571

Ultrastructural Aspects of the Developing Mammalian Zona Pellucida

J. Dietl

Introduction

During the course of oocyte development and follicular growth an amorphous material is deposited in the space between the oocyte and the surrounding granulosa cells. This material is subsequently transformed into a jelly like coat - the zona pellucida (ZP). The precise origin of the ZP and its function prior to ovulation are poorly understood (Dunbar and Wolgemuth 1984; Wolgemuth et al. 1984; Tesoriero 1984). Despite years of morphological, cytochemical, autoradiographic, and immunohistochemical studies (Chiquoine 1960; Kang 1974; van Blerkom and Motta 1979; Bousquet et al. 1981; Wolgemuth et al. 1984; Léveillé et al. 1987) the matter of the cell origin of the ZP has remained undecided. Recent work into the biochemistry of the oocyte and its zona has suggested that the ZP is secreted by the oocyte alone (Bleil and Wassarman 1980; Greve et al. 1982; Shimizu et al. 1983; Philpott et al. 1987). Other biochemical studies suggest that the ZP is the product of the granulosa cells and the oocyte (Wolgemuth et al. 1984; Maresh and Dunbar 1987). To date, biochemical evidence has not contributed toward choosing among the differing conclusions drawn from morphological and histochemical studies. No compartment of the oocyte or of adjacent follicle cells has clearly been shown to be involved in the morphogenesis of the ZP. Extracellular regulatory mechanisms are lost during in vitro studies. Therefore, such studies do not provide information about the extracellular assembly of the egg coat during oocyte growth and follicular development. More detailed biochemical, histological, and fine-structural investigation is required in order to clarify the origin of the ZP.

Pre-Zona Pellucida Stage

The oocyte of a primary, or type 3, follicle (Pedersen and Peters 1968) is generally considered to be in a "resting" state since no major morphological changes take place within the oocyte during this period. During the earliest stages of oogenesis the oocyte of the rabbit ovary is surrounded by a single layer of flat pregranulosa cells (Fig. 1). A perivitelline space does not exist. The resting primary follicle has a spheric or elliptic shape and is usually located in the superficial part of the ovarian cortical stroma where it is surrounded by dense connective tissue. The very spacious Golgi complex builds a number of separate dictyosomes (Fig. 1). These consist of five to six flattenend cisternae, which show a somewhat widened cavity on the maturing face. Surrounding the Golgi field different vesicles with an excep-

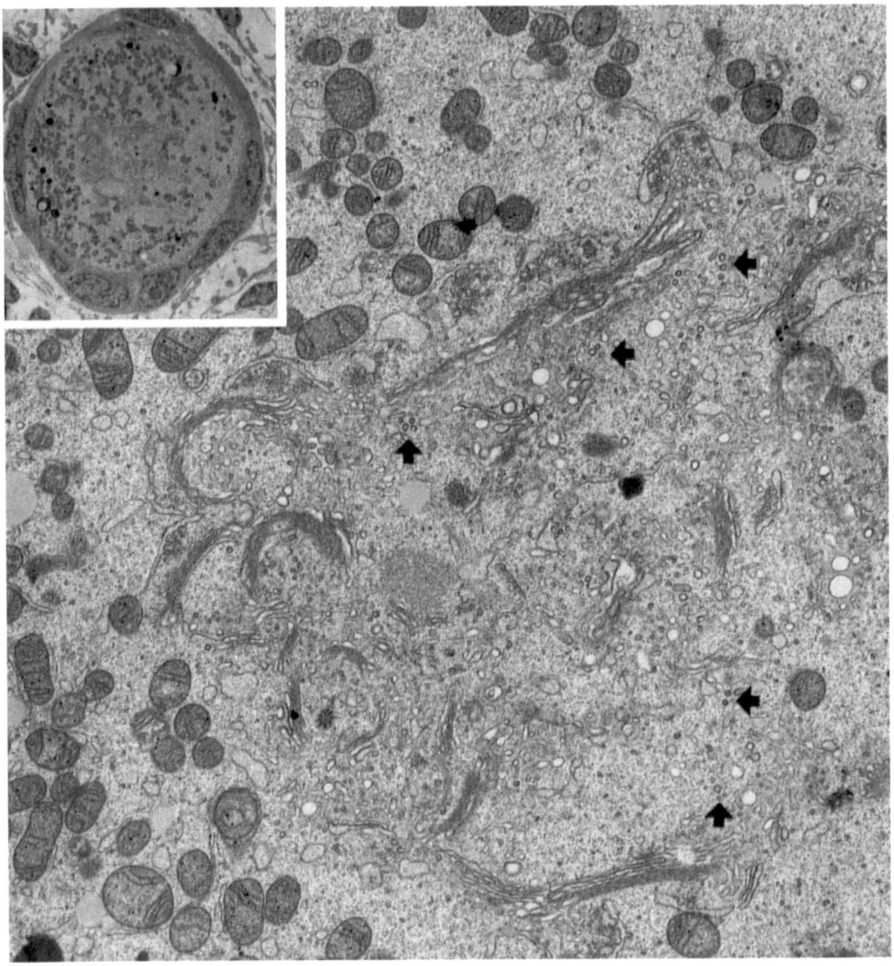

Fig. 1. Golgi apparatus of the primary follicle. At this stage an assemblage of Golgi membranes, small mitochondria, and coated vesicles *(arrows)* is still located together (× 11 000). *Inset,* primary follicle of an oocyte surrounded by a single layer of flattened pregranulosa cells (× 1260)

tional number of coated vesicles were found. In proximity to the forming face of Golgi complex cisternae were noted which probably belong to the rough endoplasmic reticulum rer), and which build numerous small vesicles. Most of the organelles in these oocytes, such as the Golgi complex, lysosomes, mitochondria, and the endoplasmic reticulum, are concentrated in a juxtanuclear position (van Blerkom and Motta 1979). This region, which is Feulgen positive, is usually crescent-shaped at the level of the light microscope and is known as the Balbiani vitelline body (Hertig 1968) or the yolk nucleus complex (Beams and Sheehan 1941). Such condensation of organelles in the juxtanuclear region of resting oocyte is considered to be the expression of the nuclear polarization of most of the oocyte metabolic activity at this stage. The follicle cells are markedly flattened and in

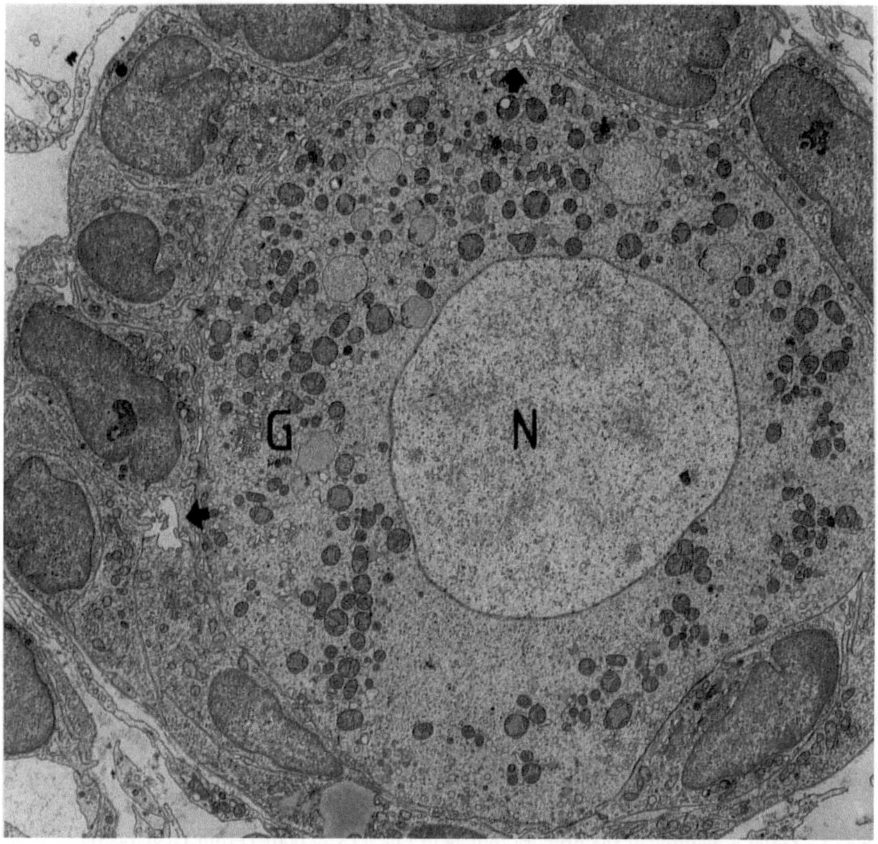

Fig. 2. Primary follicle with flat and cuboid granulosa cells. In the perivitelline space we find small pocket-shaped protrusions and cavities *(arrows)*. *N*, Nucleus; *G*, Golgi apparatus (×3600). (From Dietl 1986)

some cases may be less than a micron in thickness. The surrounding of the ovum by follicular cells may well be of some mechanical influence in that the follicular epithelium constricts the ovum. Obviously the oocyte possesses a degree of turgor during its growth period. For example, growth of the mouse oocyte results in a dramatic change in the ratio of cytoplasm to nucleoplasm, increasing from about 4:1 in the primordial oocyte to approximately 30:1 in the fully grown oocyte (Wassarman and Josefowicz 1978). At the same time, the protein content of the oocyte increases throughout growth from approximately 1 ng to about 25 ng (exclusive of zona pellucida) (Schultz and Wassarman 1977).

At a later stage of the primary follicle we find in the perivitelline space small pocket-shaped protrusions as well as a few cavities (Fig. 2). At some locations tight junctions can be recognized between the vitelline membrane and the granulosa cells, partly resembling desmosomes. Many microvilli of the oocyte membrane and some granulosa cell protrusions extend into the perioocytic space. In comparison to the previous stage the morphology of the Golgi apparatus has markedly changed (Fig. 3). During the early stages of maturation the Golgi system

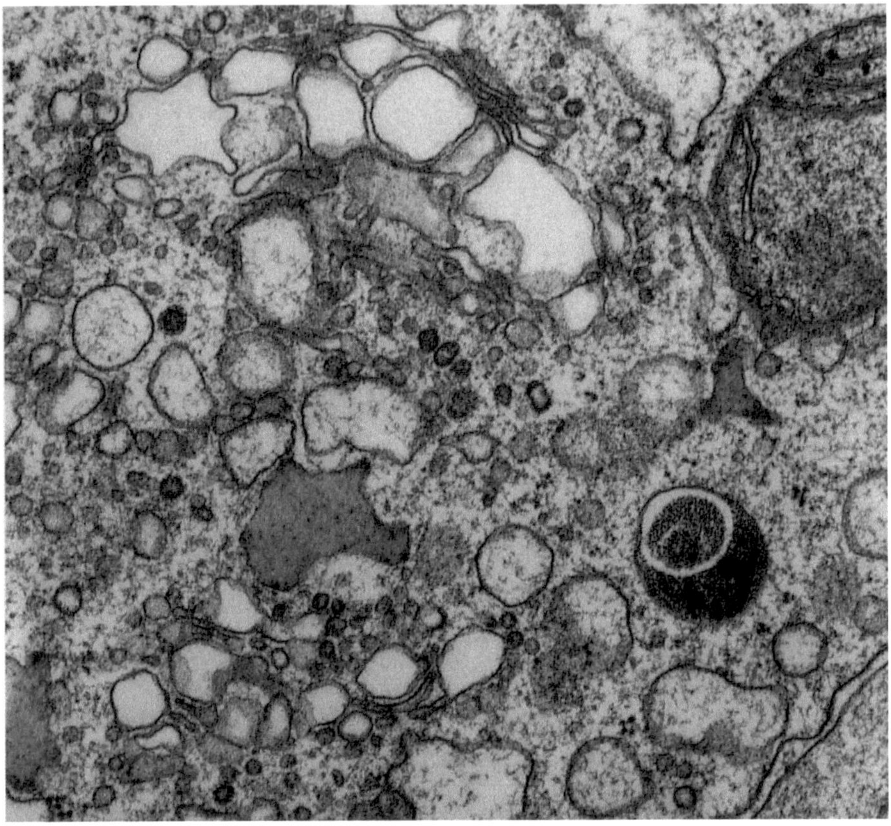

Fig. 3. Golgi apparatus of the primary follicle in Fig. 2 (× 48 000). (From Dietl 1986)

migrates from its juxtanuclear position to a subcortical location. The majority of cisternae are dilated, and the number of Golgi vesicles has increased. Larger Golgi-derived vesicles contain a filamentous material. Such vesicles appear in great number in the periphery of the oocyte. Perhaps these vesicles carry prezona material to the ooplasma membrane. The ooplasma membrane and the vesicle membrane fuse with each other in an exocytotic manner (Fig. 4).

The oocyte of the primary follicle of the mouse builds immature precursers of ZP which are accumulated in the cell (Bleil and Wassarman 1980; Shimizu et al. 1983). These immature glycoproteins show a markedly lower molecular weight than extracellular ZP glycoproteins. The maturation of ZP described above could be confirmed by Oikawa (1983) by means of density-gradient electrophoresis. This author found three ZP components in immature, yet only one in mature ovarian cells.

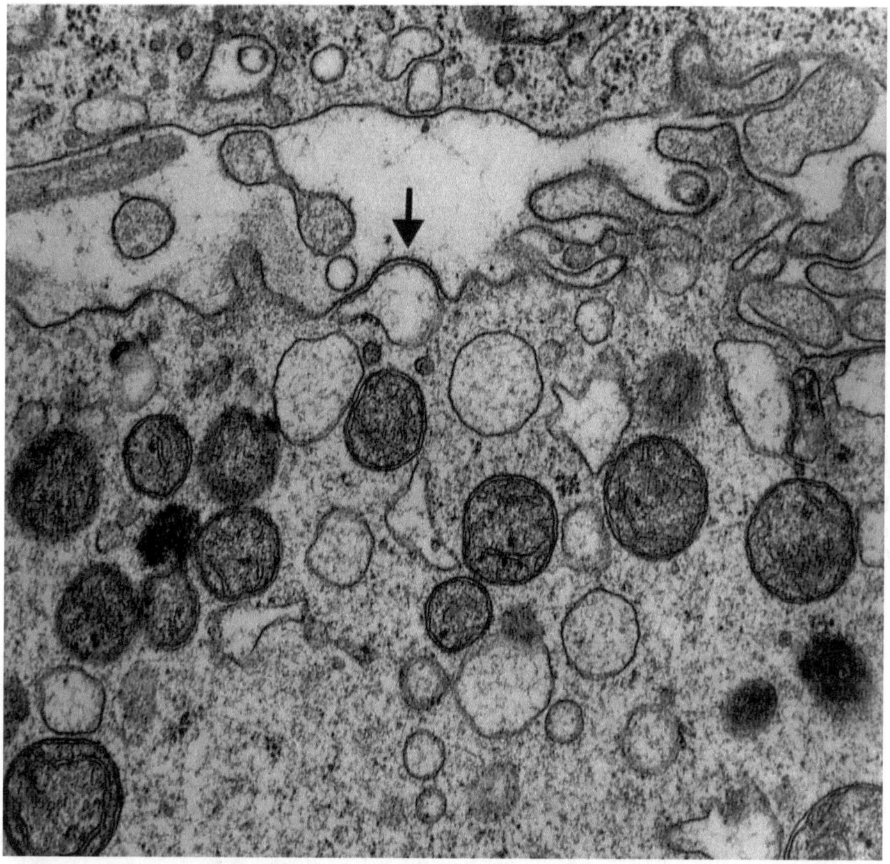

Fig. 4. Vesicles with fine filamentous material in the periphery of the ooplasma. Fusion with ooplasma membrane *(arrow)* (×44000)

Zona Pellucida Stage

The ZP is first recognizable as cloudy material lying between the pocket-shaped protrusions of microvilli of the oocyte membrane and follicular cells. At the beginning it is found only very sparsely distributed in the perivitelline space (Fig. 5). The primary appearance of ZP material coincides with an increased number of microvilli of the oocyte and cell protrusions of granulosa cells. These microvilli extend into the substance of the zona and tend to be more numerous on the surface of the ovum than on the adjacent follicle cells. Whereas the plasma membrane of the resting oocyte shows only a few short microvilli, concomitant with the formation of the ZP, the microvilli thereafter become progressively more numerous and uniformely distributed over the surface of the oocyte. ZP material is first deposited between the oocyte and a bilayer region of follicular cells (Fig. 6). The ZP ist not formed as a complete and continous extracellular membrane about the entire circumference of the oocyte but rather as discontinuous portions which fuse only secondarily to constitute a complete encasement.

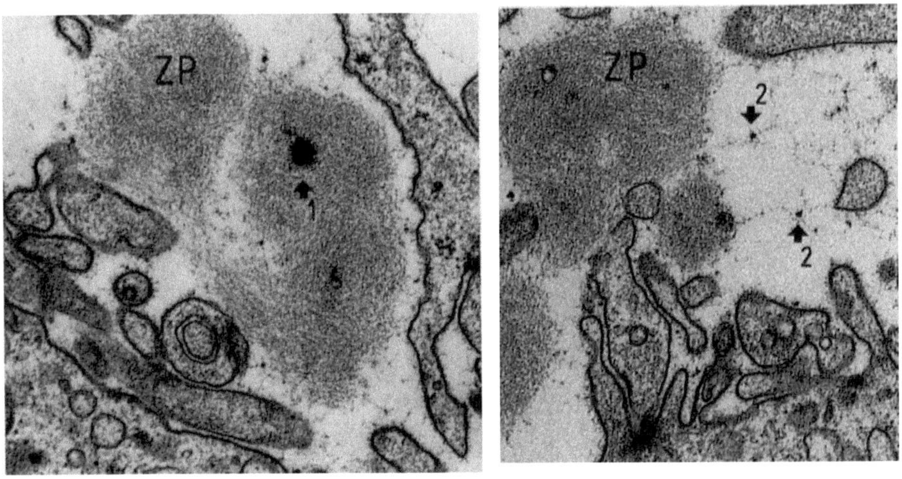

Fig. 5. First appearance of zona pellucida *(ZP)* material in the intercellular space. *Arrow 1,* granular particles with arachnoid projections; *arrows 2,* cross-linking points of ZP synthesis (× 44 000). (From Dietl 1986)

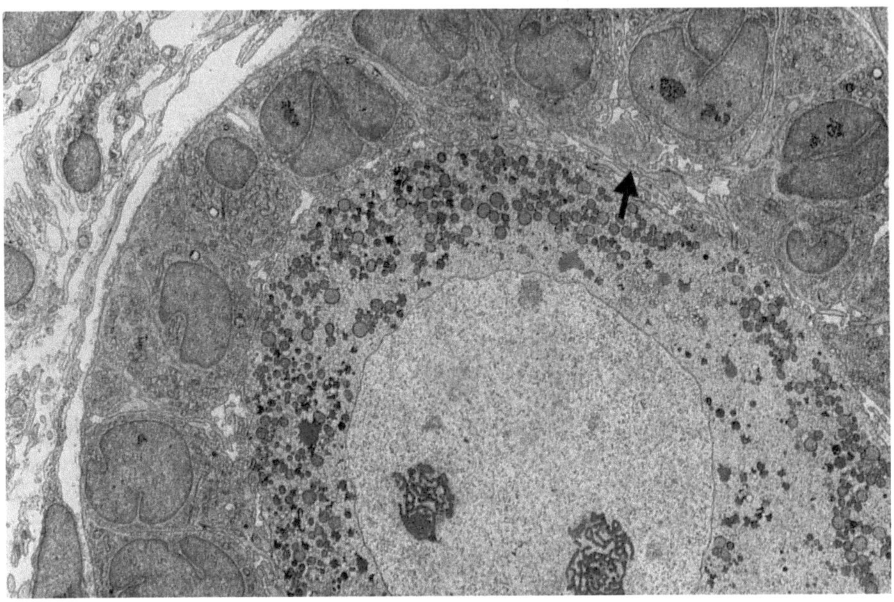

Fig. 6. Zona pellucida material is first deposited in a region of follicular cells with two layers *(arrow)* (× 3420)

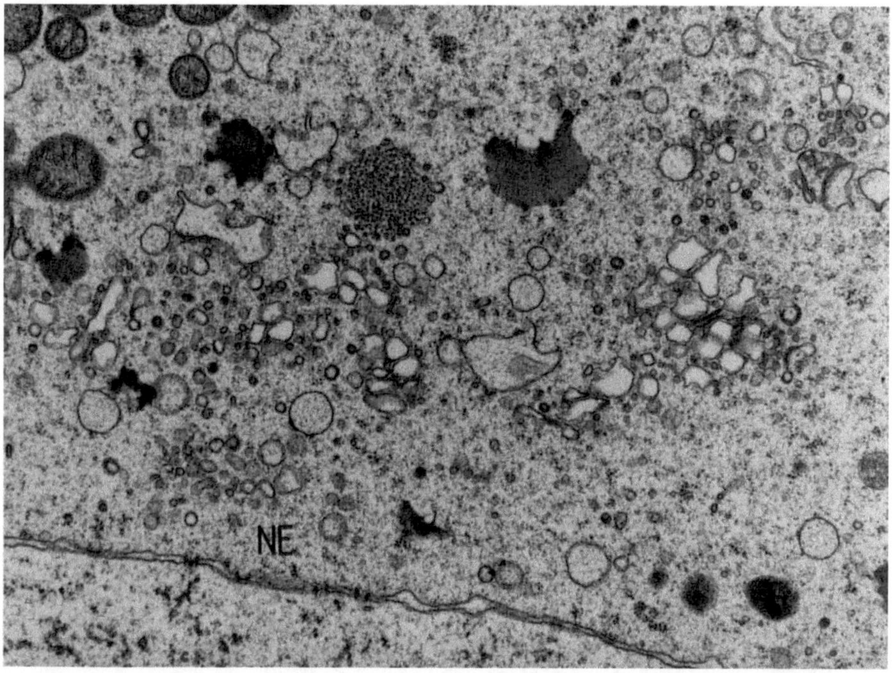

Fig. 7. Golgi vesicles at the beginning of extracellular zona pellucida synthesis in the primary follicle. *NE*, Nuclear envelope (×28000). (From Dietl 1986)

According to these morphological findings extracellular maturation of ZP starts with granules lying inside and outside of the ZP matrix (Fig. 5). Since there are fine filamentous connections to the developing ZP matrix, they seem to serve as filament branch points. Perhaps these fine granular particles are involved in the branching of ZP filaments. Greve and Wassarman (1985) could identify a glycoprotein (ZP1) of the mouse zona which has cross-linked properties with individual ZP filaments. Many studies have indicated that the ZP consists of randomly arranged, fibrillogranular material that give rise to its amorphous appearance observed in scanning-electron micrographs (Phillips and Shalgi 1980) and in transmission-electron micrographs (Wassarman and Josefowicz 1978;Dietl and Czuppon 1984). With increasing ZP growth a greater degree of polymerization and formation of the glycoprotein network obviously takes place, which has its morphologic correlate in the higher density of mature ZP fibrills. There is, however, a difference in density in the mature ZP between the inside (facing the oocyte) and the outside (facing the follicular cells): the fibrills are more densely packed inside than outside (Dietl and Czuppon 1984).

At the stage of first extracellular ZP appearance the Golgi complex (Fig. 7) is easily demonstrable. Compared with the previous stage the Golgi complex exhibits more pronounced structural changes and consists mainly of vesicular elements.

By the time that the ZP builds a closed sphere surrounding the oocyte (Fig. 8), the Golgi apparatus again appears in a different shape, consisting now mainly of

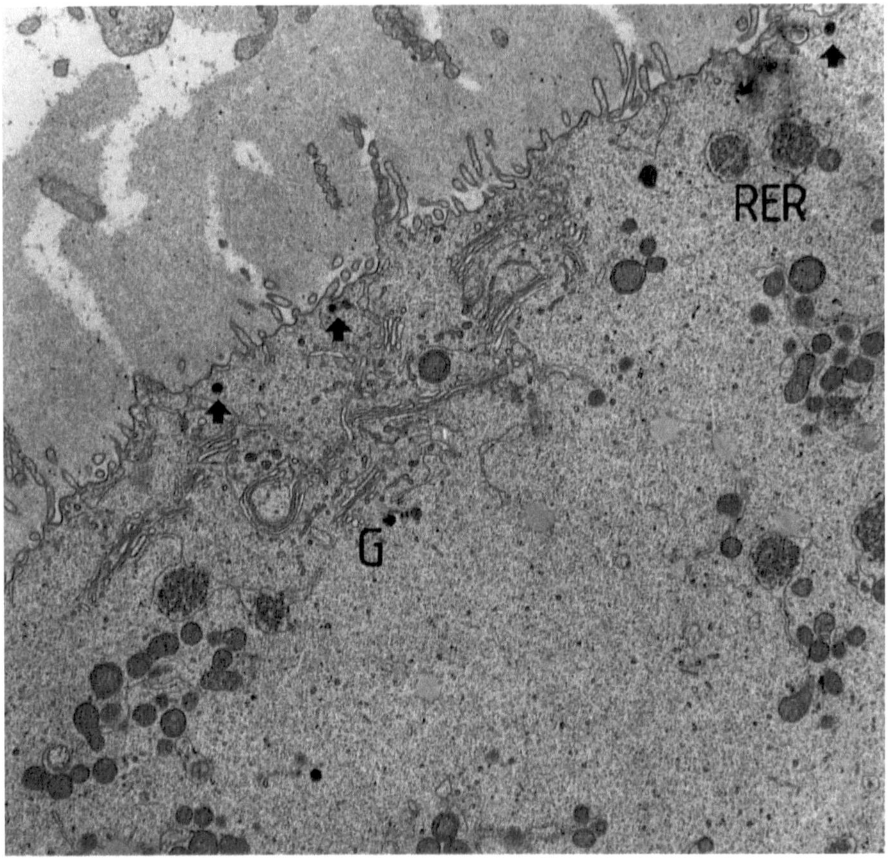

Fig. 8. Zona pellucida (thickness about 4 μm) builds a closed sphere surrounding the oocyte. *G* Golgi apparatus; *RER*, rough endoplasmic reticulum; *arrows*, cortical granules *(×10500)*. (Fron Dietl 1986)

flattened parallel cysternae lying in the periphery of the cytoplasm. On the sam side, dilated cysternae of rer are shown containing few ribosomes, and fine granu lar contents of medium electron density. In the peripheral ooplasma a higher elec tron density of cortical granules can now be detected.

We have not been able, however, to answer the question as to how the zon precursers penetrate the cytoplasmic membrane from the ooplasma to the outside An approach toward answering this question may lie in studies of the Golgi appa ratus during ZP synthesis. The Golgi complex plays a prominent role in all oocyti cellular organelles: it is highly activated during oogenesis and permanentl

Fig. 9. A Tertiary follicle. Note the many processes which arise from coronal cells and radiat through the zona pellucida (thickness, 5 μm) (×3780). **B** Electron micrograph demonstrating th typical contact between ooplasma membrane and a follicle cell process. In the process bundles o microfilaments are quite evident. Fine granules of the same morphological appearance as in Fig. are present within the zona matrix *(arrows)*. *D*, Desmosome; *O*, oocyte (×16800). (From Diet 1986)

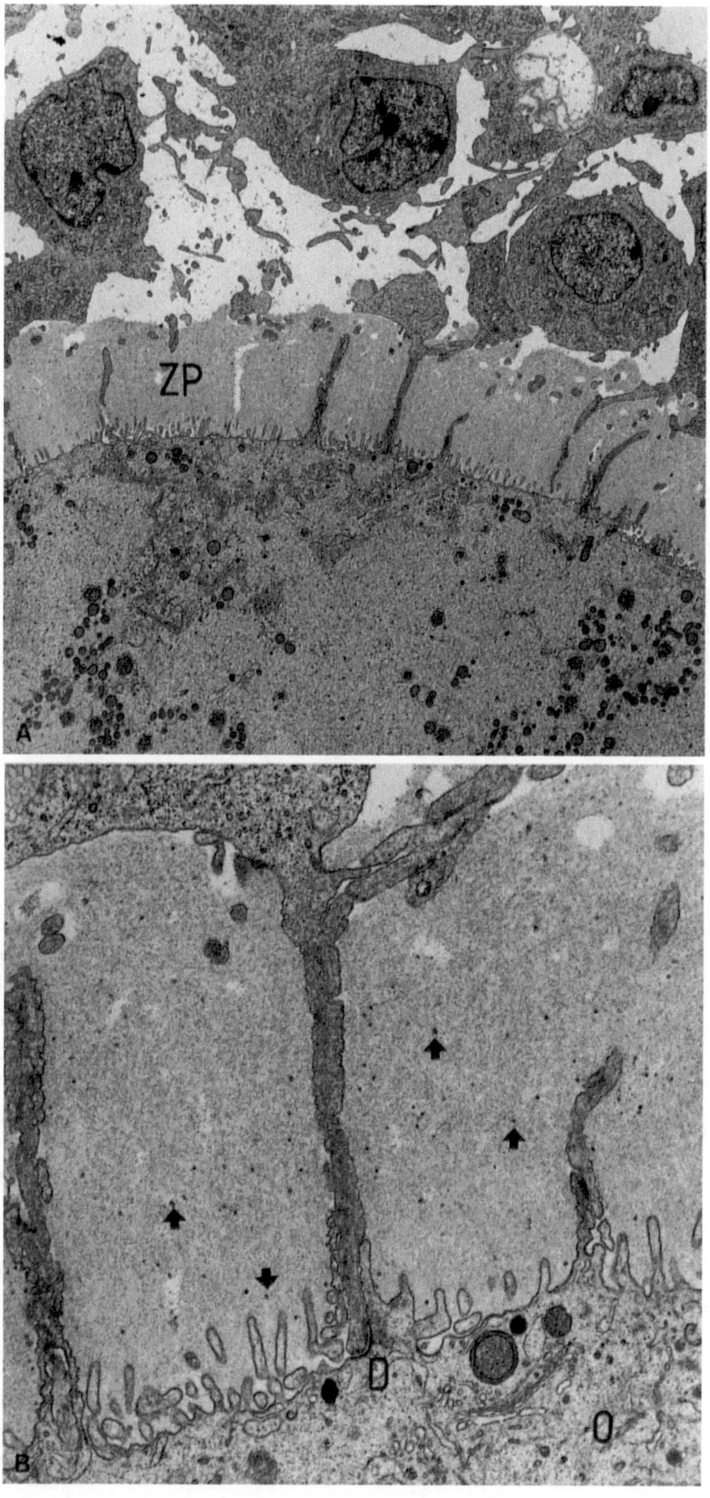
ZP
A
D
O
B

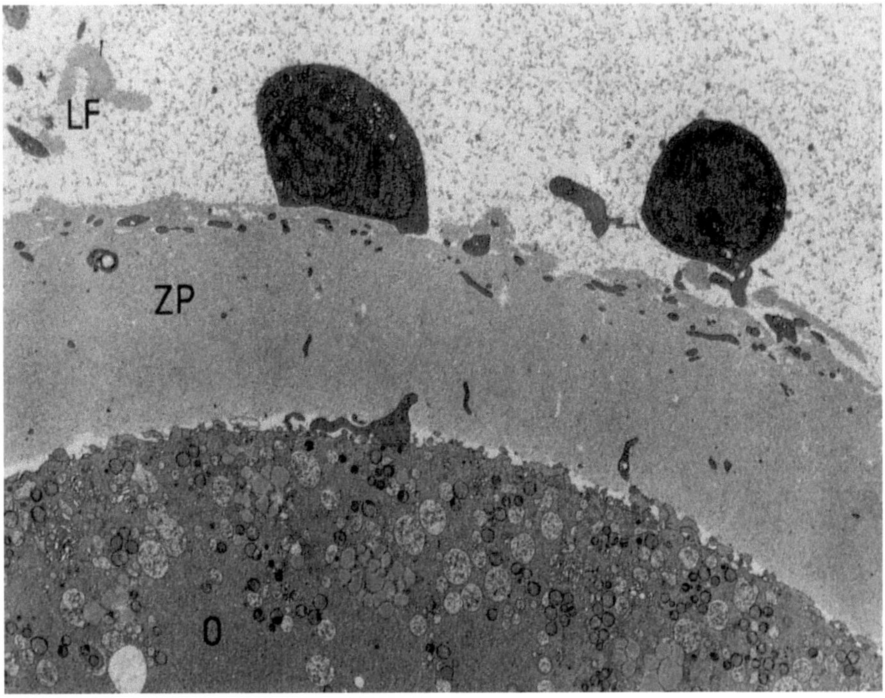

Fig. 10. Preovulatory follicle. The cumulus cells are escaped from the ZP (thickness about 8 μm). *LF*, liquor folliculi; *O*, oocyte (×4860). (From Dietl 1986)

changes its appearance. The coated vesicles described above are characteristic of the high activity of the Golgi apparatus and are most likely involved in the continual rejuvination of the Golgi complex.

The role of the Golgi apparatus in synthesis of glycoproteins, either incorporated into the cell membrane or secreted, has been investigated intensively (Michaelis and Leblond 1976; Leblond and Bennett 1977). As with glycoproteins of the nonmammalian vitelline wall, which are formed by the oocyte, it may also be supposed that this is the source of formation for ZP glycoproteins (Rosati et al. 1982). According to our present knowledge, glycoproteins are synthesized cotranslationally in the endoplasmic reticulum (ER) and are modified posttranslationally in the Golgi apparatus. They are transformed into secretory granules in the Golgi region, migrate through the cellular membrane, fuse, and are emptied exocytotically into the extracellular space (Kornfeld and Kornfeld 1980; Struck and Lennarz 1980; Rothman and Fries 1981). It seems probable that the Golgi apparatus of the oocyte plays a similar role in modifying and secreting ZP glycoproteins during oogenesis as in various other biologic systems.

During the development towards tertiary follicles the zona matrix becomes denser and wider (Fig. 9). At this stage ZP is traversed by numerous microfilaments containing channel-like cytoplasmic protrusions of the follicular cells. In the preovulatory stage we see a dissociation of cumulus cells as a result of the rapid increase of follicular fluid; the follicular cells separate gradually from the

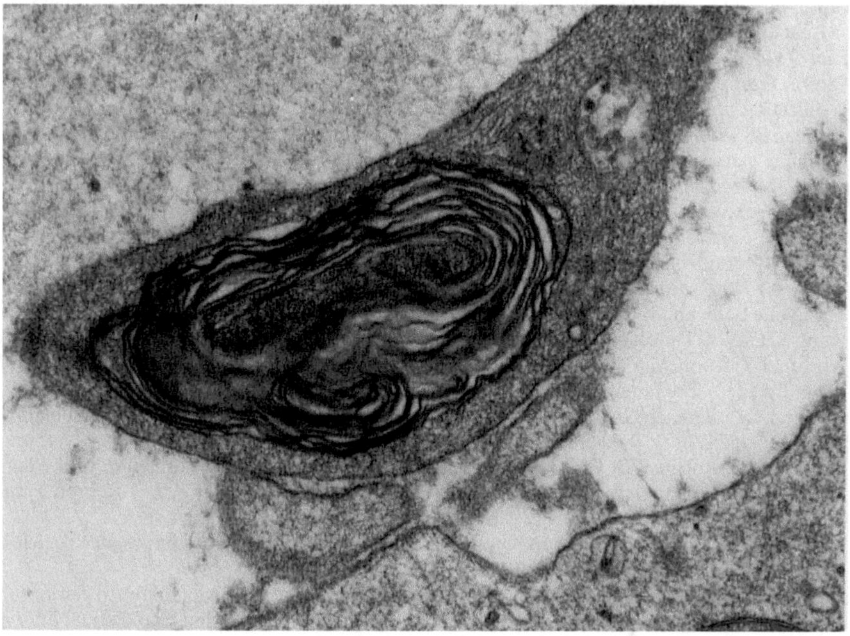

Fig. 11. Degeneration of the follicle cell process as myelin figure (×42000). (From Dietl 1986)

ZP (Fig. 10). Cell protrusions and their microfilaments retract from the ZP. The ZP has reached maturity at this stage. During the extrusion of the first polar body one can note the degeneration of follicular cell protrusions, the remnants of which are evident in the perivitelline space (Fig. 11).

Morphologic evidence supports the suggestion that the ZP is synthesized partly by the oocyte. The oocyte appears to contain the necessary intracellular organelles (Golgi, ER) in amount, location, and at the proper time to secrete a glycoprotein-like structure. This impression is, however, difficult to quantitate. It is clear from these ultrastructural studies that the formation of the ZP is a complex process. However, more detailed biochemical, histological, and ultrastructural investigation is required in order to resolve the question of the origin of the ZP.

Acknowledgement. The author would like to thank Mrs. M. Parczany for her excellent technical assistance.

References

Beams HW, Sheehan JR (1941) The yolk nucleus complex of the human ovum. Anat Rec 81: 545-554

Bleil JD, Wassarman PM (1980) Synthesis of zona pellucida proteins by denuded and follicle-enclosed mouse oocytes during culture in vitro. Proc Natl Acad Sci 77: 1029-1033

Bousquet D, Léveillé MC, Robert KD, Chapdelaine A, Bleau G (1981) The cellular origin of the zona pellucida antigen in the human and hamster. J Exp Zool 215: 215-218

Chiquoine AD (1960) The development of the zona pellucida of the mammalian ovum. Am J Anat 106: 149-196
Dietl J (1986) Struktur und Funktion der Zona pellucida. Enke Verlag, Stuttgart
Dietl J, Czuppon AB (1984) Ultrastructural studies of the porcine zone pellucida during the solubilization process by Li-3.5-dijodosalicylate. Gamete Res 9: 45-54
Dunbar BS, Wolgemuth DJ (1984) Structure and function of the mammalian zona pellucida, a unique extracellular matrix. In: Satir BH (ed) Modern cell biology. New York, pp 77-111
Greve JM, Wassarman PR (1985) Mouse egg extracellular coat is a matrix of interconnected filaments posessing a structural repeat. J Mol Biol 181: 253-264
Greve JR, Salzmann GS, Roller RJ, Wassarman PR (1982) Biosynthesis of the major zona pellucida glycoprotein secreted by oocytes during mammalian oogenesis. Cell 31: 749-759
Hertig AT (1968) The primary human oocyte. Some observations of the fine structure of Balbianis vitelline body and the origin of the annulate lamellae. Am J Anat 122: 107-138
Kang YH (1974) Development of the zona pellucida in the rat oocyte. Am J Anat 139: 535-566
Kornfeld R, Kornfeld S (1980) Structure of glycoproteins and their oligosaccharide units. In: Lennarz WJ (ed) The biochemistry of glycoproteins and proteoglycans. Plenum, New York
Leblond CP, Bennett G (1977) Role of the Golgi-apparatus in terminal glycosylation. In: Cell biology 1976-1977. Rockefeller University Press, New York, pp 326-336
Leveille MC, Robert KD, Chevalier S, Chapdelaine A, Bleu G (1987) Formation of the hamster zona pellucida in relation to ovarian differentiation and follicular growth. J Reprod Fertil 79: 173-183
Maresh GA, Dunbar BS (1987) Expression of specific proteins by ovarian primary follicles cultured in vitro. Biol Reprod [Suppl 1] 36: 88a
Michaelis J, Leblond CP (1976) Transport of glycoprotein from Golgi-apparatus to cell surface by means of carrier vesicles as shown by radioautography of mouse colonic epithelium after injection of tritiated fucose. J Microsc Biol Cell 25: 243-248
Oikawa T (1983) Detection of differences between components of immature and mature zonae pellucidae by the method of density gradient pag-electrophoresis. J Reprod Immunol [Suppl]: 5: 25
Pedersen T, Peters H (1968) Proposal for the classification of oocyte and follicles in the mouse ovary. J Reprod Fertil 17: 555-557
Phillips DM, Shalgi RM (1980) Surface properties of the zona pellucida. J Exp Zool 213: 1-8
Philpott CC, Ringuette MJ, Dean J (1987) Oocyte specific expression and developmental regulation of ZP 3, the sperm receptor of the mouse zona pellucida. Dev Biol 121: 568-575
Rosati F, Cotelli F, DeSantis R, Monroy A, Pinto MR (1982) Synthesis of fucosyl-containing glycoproteins of the vitelline coat in oocytes of *Ciona intestinalis* (Ascidia). Proc Natl Acad Sci 79: 1908-1911
Rothmann JE, Fries E (1981) Transport of newly synthesized vesicular stomatitis viral glycoprotein to purified Golgi-membranes. J Cell Biol 89: 162-168
Schultz RM, Wassarman PM (1977) Biochemical studies of mammalian oogenesis: protein synthesis during oocyte growth and miotic maturation in the mouse. J Cell Sci 24: 167-194
Shimizu S, Masahiro T, Dean J (1983) In vitro biosynthesis of three sulfated glycoproteins of murine zona pellucida by oocytes grown in follicle culture. J Biol Chem 258: 5858-5863
Struck OK, Lennarz WJ (1980) The function of saccharidelipids in synthesis of glycoproteins. In: Lennarz WJ (ed) The biochemistry of glycoproteins and proteoglycans. Plenum, New York, pp 34-83
Tesoriero JV (1984) Comparative cytochemistry of the development ovarian follicles of the dog, rabbit, and mouse: origin of the zona pellucida. Gamete Res 10: 301-318
Van Blerkom J, Motta P (1979) The cellular basis of mammalian reproduction. Urban and Schwarzenberg, Münich
Wassarman PM, Josefowicz WJ (1978) Oocyte development in the mouse: an ultrastructural comparison of oocytes isolated at various stages of growth and miotic competence. J Morphol 156: 209-236
Wolgemuth DJ, Celenza J, Bundman DS, Dunbar BS (1984) Formation of the rabbit zona pellucida and its relationship to ovarian follicular development. Dev Biol 106: 1-14

Zona Binding Sites of the Spermatozoon

R. B. L. Gwatkin

Introduction

Binding of the spermatozoon to the zona pellucida of the egg is one of the essential steps in mammalian fertilization. If we knew the identity of the responsible molecules and their active sites, we should be able to diagnose infertility involving defects at these sites. We could also block them to achieve contraception without the need to interfere with gametogenesis or to interrupt development after conception.

Using sea urchins, from which large numbers of gametes can be easily obtained, Vacquier and Moy (1977) isolated an essentially carbohydrate-free protein receptor, bindin, from the acrosomal processes of the sperm. The complete sequence of bindin was reported recently (Gao et al. 1986). The complementary proteoglycan receptor on the uncrosslinked fertilization envelope of the sea urchin egg has also just been isolated and partially characterized (Ruiz-Bravo et al. 1986).

In mammals, from which relatively fewer gametes can be obtained and with greater difficulty, considerable advances have nevertheless been made over the past decade in our knowledge of gamete receptors. In this chapter the current status of our knowledge of the zona binding sites in the mammalian spermatozoon will be presented. Since the sperm receptor-for-zona cannot be discussed apart from its complimentary receptor-for-sperm the latter will also be covered briefly. In fact, we shall begin with this receptor since it was the first to be investigated.

Receptor-for-sperm on Egg

A decade ago it was reported that heat-solubilized hamster zonae inhibited the binding of hamster spermatozoa to eggs in vitro (Gwatkin and Williams 1977). This action was species-specific. Solubilized zonae from eggs that had undergone a zona reaction were inactive. Thus, a receptor-for-sperm had been demonstrated for the first time. These authors suggested that the receptor might be a glycoprotein. Later it was shown that homologous solubilized zonae blocked the binding of capacitated, but not of noncapacitated, hamster and bovine spermatozoa to their respective eggs (Gwatkin and Williams 1978).

Shortly thereafter, the three major glycoproteins of the mouse zona were isolated, and it was shown that one of these, ZP3, carried the receptor (Bleil and Wassarman 1980). Following proteolytic digestion of ZP3, glycopeptide subunits of molecular weight 1.5-6-kd were found to bear the active sites (Florman et al.

1984). These were inactivated by chemical removal of the O-linked, but not the N-linked, carbohydrate side chains of ZP3 (Florman and Wassarman 1985). Clones coding for mouse ZP3 have been isolated recently, and it has been shown that the ZP3 gene is relatively hypomethylated in ovarian tissue, where it is expressed, compared to other tissues where it is not (Ringuette et al. 1986).

Concurrent investigations of the pig zona have shown that its receptor-for-sperm is located in a 38-kd HPLC fraction (Dietl et al 1983) and in a 58-kd ion-exchange column fraction (Sacco et al. 1984). Subunit isolation has not yet been reported.

Sensitivity to Carbohydrates and Enzymes

One approach which has provided information as to the importance of the carbohydrate moiety is to determine the sensitivity of the receptor to various glycosidases and to competitive inhibition by various saccharides. The mouse zona receptor was shown to be sensitive to β-*N*-acetylglucosaminidase (Shur and Hall 1982a). The sperm binding sites of the hamster zona are not sensitive to neuraminidase, β-glucosidase, or β-galactosidase (Gwatkin and Anderson 1973), but they are inactivated by *N*-acetylglucosaminidase (Ahuja 1982).

N-Acetylglucosamine is the predominant saccharide of the porcine zona (Gwatkin 1981; Henderson et al. 1987) and of its receptor-bearing subunit (Sacco et al. 1986). The pig zona also contains significant amounts of *N*-acetylgalactosamine, galactose, mannose, and fucose, hence one might anticipate that one or more of these sugars, or their polymers, would compete for sperm binding. No studies appear to have been done with pig gametes. However, fucoidin (an algal polysaccharide, rich in fucose) inhibits the binding of guinea pig sperm to the zona (Huang et al. 1982). Fucoidin and α-methyl mannoside inhibit the union of rat gametes (Shalgi et al. 1986). Glycoproteins rich in *N*-acetylglucosamine, such as orosomucoid and fetuin, when desialylated inhibit the binding of hamster gametes (Ahuja 1982).

It thus appears that the receptor-for-sperm on the zona depends on essential terminal carbohydrates. As we shall see, these may bind to the complimentary receptors on the sperm, which can either be a lectin or a glycosyl transferase.

Zona-Associated Proteins

Factors which must be taken into consideration when investigating the origin and location of the sperm-binding molecules of the zona are its porous structure and its tendency to incorporate extraneous materials. Fox and Shivers (1975) were the first to show that oviductal antigens can become associated with the hamster zona after ovulation. Léveillé et al. (1987) subsequently established that antisera pre

Fig. 1 A–C. Sections through vitellus and zona. **A** Hamster ovulated egg exposed to rabbi antiovulated hamster zona serum (diluted 1:1000). **B** Ovarian egg exposed to same antiserur (diluted 1:50). **C** Ovulated egg exposed to normal rabbit serum. All sections were subsequentl exposed to 10-nm protein A-gold particles. Note the heavy labeling *(black dots)* of the zona of th ovulated egg **A** and the sparser labeling of the ovarian zona at a 20-fold higher antiserum cor centration **B.** The control **C** was unlabeled. *V*, vitellus; *PVS*, pervitelline space; *ZP*, zona pellu cida

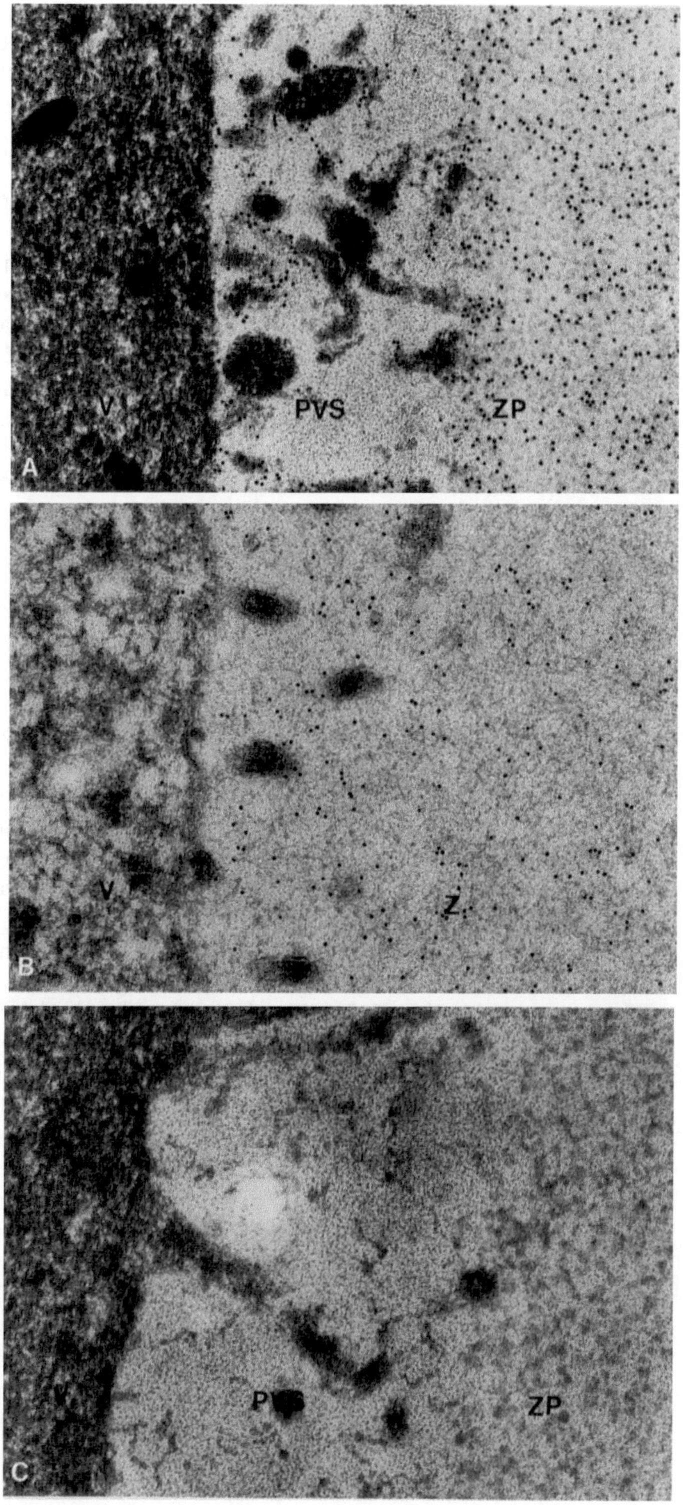
V
PVS
ZP
A
V
Z
B
P
ZP
C

pared against the zonae of ovulated hamster eggs cross-reacted with the cells o the oviduct fimbria, ampulla, and isthmus. After dispersion of the cumulus oopho rus from preovulatory hamster oocytes with hyaluronidase, the pores left in th outer portion of the zona by withdrawal of the corona cell processes were foun to contain extracellular material that could be visualized by electronmicroscop after staining with ruthenium red (Talbot 1984). Incorporation of extracellular gly cosaminoglycans from the cumulus matrix has been detected in mouse zona shortly after fertilization (Fowler et al. 1986). The zonae of ovulated porcine egg were found to possess two oviductally derived glycoproteins that were absent fron those of follicular oocytes (Brown and Cheng 1986). An antiserum to the isolate zonae of ovulated hamster eggs reacted strongly with the zonae of these eggs at dilution of 1:1000, but produced only a weak labeling at 1:50 against the zona of ovarian oocytes (Fig. 1). This suggests that a major antigenic component of th ovulated zona is not present before ovulation. It also seems probable that th pores on the inner aspect of the zona will be found to contain pervitelline mate rial. To what extent do these extraneous materials contribute to specific zona func tions? Future research should be directed to answering this question.

Receptor-for-Zona on the Spermatozoon

Sperm Antigens

One approach to identifying this receptor on the sperm is to develop antibodie against specific sperm antigens and then to determine their effect on gamete binc ing. A monoclonal antibody to the PH-20 antigen, isolated from guinea pig spe matozoa, has been shown to block sperm binding to the zona (Primakoff et a 1985). PH-20 is a curious antigen because it appears to migrate from the surface the posterior head of the sperm to its inner acrosomal membrane during the acr some reaction (Myles and Primakoff 1984).

Several other antibodies have been produced against sperm plasma membrar antigens. Some of them block fertilization, although it is not known that they c so by preventing binding. O'Rand et al. (1984) have raised monoclonal antibodi to the rabbit sperm glycoprotein, RSA1, found on spermatogenic cells as well on spermatozoa. When rabbit sperm were preincubated with these antibodie their ability to fertilize eggs was blocked. Polyclonal autoantibodies to guinea p sperm have been shown to prevent sperm from binding to the zona (Yanagimac et al. 1981). Naz et al. (1986) isolated a 23-kd fertilization antigen, FA-1, fro mouse and human testicular spermatozoa and developed a monoclonal antiboc against it which inhibited the fertilization of mouse oocytes in vitro. In some of tl above reports sperm binding may have been prevented, but in other instanc another fertilization mechanism may be responsible. Thus M42, a monoclon antibody directed at the plasma membrane of the mouse sperm, was shown to pı vent fertilization by inhibition of the sperm acrosome reaction (Saling 1986).

A major problem in using antibodies to define binding sites is that they c also act indirectly by blocking access to the receptor rather than reacting with tl receptor itself. Other approaches (see below) have been more productive.

Sperm Membrane Fractions with High Affinity for Zona

Another way of defining the zona binding sites has been to isolate sperm plasma membrane fractions and to determine which of these fractions competitively blocks binding of the sperm to eggs in vitro (Petersen et al. 1980). Isolation was achieved by releasing membranes from cauda epididymal pig sperm, then separating the active fractions by chromatography on dextran sulfate columns, followed by further fractionation on an affinity column of isolated zonae (Peterson et al. 1985). Several active fractions were obtained with molecular weights of approximately 30-, 45-, and 70-kd. Caput sperm, which do not bind to oocytes, do not contain these proteins (Peterson et al. 1986).

Probing Western Blots

Yet another approach to identifying the receptor-for-zona on the sperm has been to probe Western blots of electrophoretically separated sperm components with solubilized zonae in order to determine which components exhibit zona binding. Sullivan and Bleau (1985) extracted hamster sperm with the detergent Nonidet P-40, which removed the sperm plasma membrane and acrosomal matrix. They then submitted the extracts to polycrylamide gel electrophoresis with sodium dodecyl sulfate (SDS-PAGE) and transferred the electrophoretically separated proteins to nitrocellulose. These blots were then exposed to ^{125}I-solubilized zonae. The labeled zonae bound specifically to a 26-kd glycoprotein not present in the sperm of several other species. This glycoprotein was presumed to be from the sperm plasma membrane.

O'Rand et al. (1985) used a similar approach with rabbit, pig, human, and mouse sperm, finding several different polypeptides that bound ^{125}I-solubilized zona in each species, with major differences between species. Another study was reported by Brown and Jones (1987), using deoxycholate extracts of pig sperm. Consistent binding was obtained to a 53-kd sperm component, with weaker binding to 18-, 38-, and 67-kd proteins. Two seminal plasma components (19–24 and 65-kd) were also bound. The data of Brown and Jones differ from those of O'Rand et al. (1985), who found strong binding to 16- and 18-kd sperm components, with lesser binding to those of 45-, 53-, 63-, and 88-kd. It is apparent that this approach is uncovering a variety of zona-binding components. However, it is unclear at present how this binding relates to the physiology of fertilization. It seems unlikely that they are all involved in sperm binding. We also do not know where the binding components are located, whether in the plasma membrane, in the acrosomal matrix, which is also extracted (e.g., see Sullivan and Bleau 1985, Fig. 4A), or elsewhere.

Labeled Zonae

A more direct approach to locating zona binding sites is to treat sperm with solubilized zonae, or a subunit. Binding sites for zona on the spermatozoon are then recognized by prelabeling or by a second antibody. In early studies Gwatkin (1978) added heat-solubilized hamster zonae to a suspension of capacitated ham-

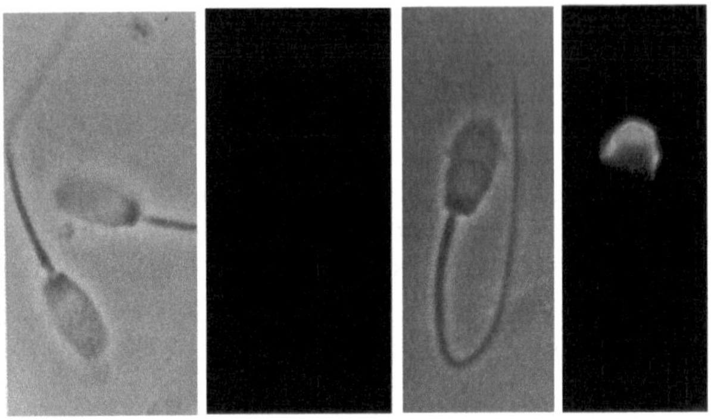

Fig. 2. Porcine sperm after acrosome reaction was induced by incubation for 45 min with 5-μM ionophore A23187. After 15 min *p*-aminobenzamidine was added to prevent loss of the partly dispersed acrosomal matrix. The sperm were then fixed in carbodiimide and exposed either to buffered saline (control) or to buffered saline and heat-solubilized zonae (0.5 mg/ml). Zona binding sites on the sperm were located by rabbit anti-zona serum followed by fluorescein-conjugated goat anti-rabbit globulin. *Lighter panels* are phase-contrast photograph, and *darker panel* is by fluorescence. Note that control sperm *(left)* are unlabeled, while expanded acrosomal matrix is intensely labeled with a fainter labeling of the postacrosomal region

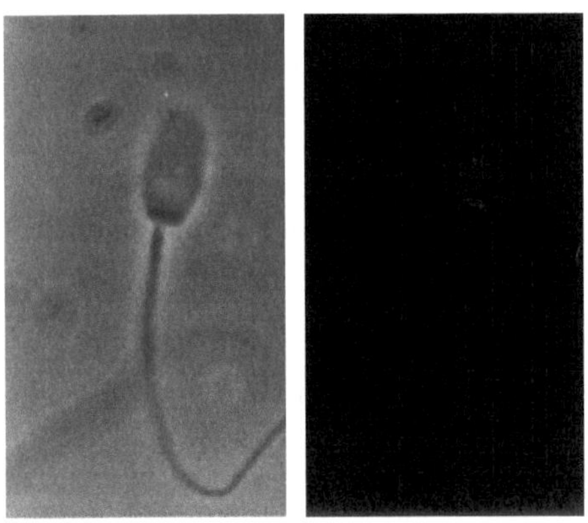

Fig. 3. Same conditions as in Fig. 2, but incubation was continued for 60 min before *p*-aminobenzamidine was added. Phase contrast photograph on *left*, fluorescence on *right*. The postacrosomal plasma membrane fluorescence is still present, but the acrosomal fluorescence has disappeared. Loss of the acrosomal matrix has exposed the inner acrosomal membrane which does not bind zona

ster spermatozoa containing acrosome-reacted and non-acrosome-reacted sperm. The sperm were then centrifuged through a stable discontinuous gradient containing rabbit anti-hamster zona serum and fluorescein-conjugated goat anti-rabbit IgG, separated by washing buffer. When the sperm were examined under a fluorescence microscope, labeling was found over the entire plasma membrane of the head and tail before the acrosome reaction. After the reaction and loss of the acrosome matrix, however, the exposed inner acrosomal membrane was not labeled. In similar experiments with porcine sperm (Gwatkin and Klein 1985) the inner acrosomal membrane also failed to bind solubilized zonae (Fig. 3). When, following the acrosome reaction, loss of the acrosome matrix was prevented by the addition of a protease inhibitor, *p*-aminobenzamidene, (Green 1978; Fraser 1982), the matrix was strongly labeled (Fig. 2).

Recently O'Rand and Fisher (1987) labeled rabbit zonae with fluorescein, then heat-solubilized them, and applied the labeled solution to live sperm. Before the acrosome reaction the label was found over the entire head and midpiece or as patches along the apical ridge. After the acrosome reaction labeling on the head became restricted to the postacrosomal region of the midpiece.

Labeled Subunits of the Zona Pellucida

Bleil and Wassarman (1986) iodinated the mouse zona's sperm receptor, ZP3, and studied its binding to whole sperm by autoradiography. ^{125}I-labeled ZP3 was found to bind to the heads of those sperm that had not yet undergone an acrosome reaction. Reacted sperm were not labeled. However, under these conditions an early phase of the acrosome reaction, i.e., membrane vesiculation, would not have been detected and after the acrosome reaction the acrosome matrix would have been lost. Iodinated ZP2, the major glycoprotein of the mouse zona that was not previously thought to have sperm binding sites, bound preferentially to acrosome-reacted sperm. The authors suggested that it could serve as a secondary receptor.

Role of Saccharides

We have already observed that a major saccharide component of the zona glycoproteins is *N*-acetylglucosamine. Shur and Hall (1982b) have reported convincing evidence that sperm surface galactosyltransferase may add galactose residues to *N*-acetylglucosamine on the zona, thereby binding the gametes together. UDP-dialdehyde was found to inhibit sperm binding to the mouse zona and sperm surface galactosyltransferase activity to identical degrees. On uncapacitated spermatozoa, the transferases are apparently blocked by polylactosaminyl residues (Shur and Hall 1982b). Their release may be responsible for sperm capacitation. Capacitation of hamster spermatozoa by cumulus cells in vitro was shown many years ago to involve the action of glycosidases (Gwatkin and Anderson 1973).

Status of the Sperm Acrosome During Binding

Whether the acrosome reaction occurs before or after binding of the sperm to the zona appears to depend on the species. It has been reported to occur before binding with guinea pig (Huang et al. 1981) and rabbit (Kuzan et al. 1984) gametes. In other species, such as the mouse, rat, and hamster, it apparently occurs after binding. Storey et al. (1984) have shown that mouse spermatozoa binding to the zona have intact acrosomes, as judged by chlortetracycline prelabeling (Saling and Storey 1979). In the initial binding step the impermeability of the plasma membrane to small ions was retained, as shown by an aminoacridine-fluorescent pH probe (Lee and Storey 1985). Consistent with these studies is the recent report of Endo et al. (1987), who showed that phorbol esters block the acrosome reaction of mouse sperm without affecting binding. In the rat as well it appears that sperm arrive at the zona pellucida with substantially intact acrosomes (Phillips and Shalgi 1982).

With hamster gametes what may be an early phase of the acrosome reaction has been detected as the sperm traverse the cumulus oophorus (Cummins and Yanagimachi 1986). The plasma and outer acrosomal membranes bulge, and the acrosomal matrix becomes less uniform. However, loss of the acrosome cap does not occur until after the sperm reach the zona. Sperm that have undergone this final step in the acrosome reaction are infertile (Gwatkin et al. 1976).

It has been demonstrated that a fucose-binding protein previously isolated from pig sperm (Toepfer-Petersen et al., 1985), becomes progressively exposed during the course of an ionophore-induced acrosome reaction (Friess et al. 1987a). The label was located in the acrosome and outer acrosomal membrane, but not on the inner acrosomal membrane (Friess et al. 1987b). An identical labeling was obtained with solubilized labeled zonae (E. Toepfer-Petersen, personal communication), suggesting that the fucose-binding protein may be a sperm receptor located in the acrosome. Since whole zona were used for labeling, it was not known which of the zona components was involved in binding.

When binding occurs before the acrosome reaction, it is presumably to the sperm plasma membrane. After the acrosome reaction binding might be to the plasma membrane over the postacrosomal region or to the acrosomal matrix, some portion of which is probably retained until the penetration of the zona is completed. Binding to the acrosomal matrix could also constitute a secondary binding process following acrosome reaction on the zona pellucida. Present evidence is that the inner acrosomal membrane is not involved, since it does not bind solubilized zona (Gwatkin 1978; O'Rand and Fisher 1987).

A likely candidate for the binding material in the acrosomal matrix is acrosin, i.e., binding would occur as an enzyme-substrate reaction following the conversion of proacrosin to acrosin. Recent studies with pig gametes in fact support a role for the acrosin system in sperm-egg binding. Thus it has been shown that ^{125}J-pig zona binds preferentially to sperm acrosin (Jones et al. 1988), apparently to a nonenzymatic site (Töpfer-Petersen and Henschen 1988).

The liberation of soluble acrosin during the ram acrosome reaction suggests that proacrosin is not bound to the inner acrosomal membrane but is a component of the acrosomal matrix (Huneau et al. 1984). The sensitivity of sperm binding to trypsin inhibitors (Saling 1981) may therefore be evidence that the acrosomal matrix plays a role in the binding process. However, Benau and Storey (1987) suppose that this "trypsin" is located in the sperm plasma membrane. They provide evidence that it differs from trypsin and acrosin in that it does not hydrolyze lysyl and arginyl trypsin substrates, although it hydrolyses the serine protease active-site titrant 4-methyl-umbelliferyl *p*-guanidinobenzoate. Clearly, more studies will be required to solve this puzzle.

Induction of the Acrosome Reaction by the Zona

Bleil and Wassarman (1983) were the first to report that the solubilized zona pellucida of the mouse egg induces the acrosome reaction of capacitated mouse sperm

Table 1. Binding of hamster sperm and acrosome caps to zonae pellucidae and to whole eggs

Whole sperm or Subunits	Mean number per		Significance of difference
	Zona	Egg	
Motile sperm with intact acrosomes	2.43 ± 0.36	0.50 ± 0.09	Highly significant ($p < 0.005$)
Motile sperm undergoing acrosome reaction (caps partly detached)	0.37 ± 0.03	1.40 ± 0.13	Highly significant ($p < 0.001$)
Motile sperm that have completed acrosome reaction (caps lost)	3.67 ± 0.53	2.37 ± 0.34	Insignificant ($p > 0.1$)
Detached caps	4.67 ± 0.87	3.60 ± 0.35	Insignificant ($p > 0.4$)

Data are derived from six experiments, each employing ten eggs and zonae ± s.e. Washed cumulus-free hamster eggs and isolated zonae were added to microdrops of capacitated sperm. After incubation for 30-40 min eggs and zonae were washed and the number of motile sperm with intact acrosomes, those with partly detached acrosome caps, and those without acrosomes were counted under phase contrast optics. The number of detached caps was also recorded.

They also showed that this activity was attributable to ZP3, the same zona glycoprotein that carries the receptor-for-sperm. Solubilized homologous zonae have also been shown to induce the rabbit sperm acrosome reaction (O'Rand and Fisher 1987).

More recently Cherr et al. (1986), using high-speed videomicrography, followed the final loss of the hamster sperm acrosome cap as capacitated hamster sperm traversed the cumulus oophorus and encountered cumulus-free eggs in a fertilization chamber. They established that a higher proportion of the sperm completed their acrosome reaction on the zona surface when the cumulus was present than when it was absent. Thus, the cumulus may act synergistically with the zona to induce rapid onset of the acrosome reaction. Our laboratory has observed that more sperm undergo an acrosome reaction on the surface of the whole egg than on the surface of the isolated zona after coincubation of the gametes for 30-40 min (Table 1). In this case it would be the material from the vitellus or perivitelline space that would act synergistically with zona to induce the acrosome reaction. Possibly one or more glycosaminoglycans may be involved. They are known to be present in the cumulus matrix (Ball et al. 1982). Glycosaminoglycans, such as heparin, chondroitin sulfate, and hyaluronic acid, are known to stimulate the hamster (Meizel and Turner 1985) and rabbit (Lenz et al. 1983) acrosome reaction.

No penetration of isolated zonae by sperm was observed up to 80 min after contact, but when the hamster gametes were examined by scanning electron microscopy at 160 min, penetration was apparent (Fig. 4). Thus it is clear that the vitelline factor accelerates the acrosome reaction but is not essential for it to occur.

Binding of Sperm to Isolated Zonae and Whole Eggs

The apparent need for vitelline and perivitelline material to accelerate the acrosome reaction may help to explain some puzzling observations made many years

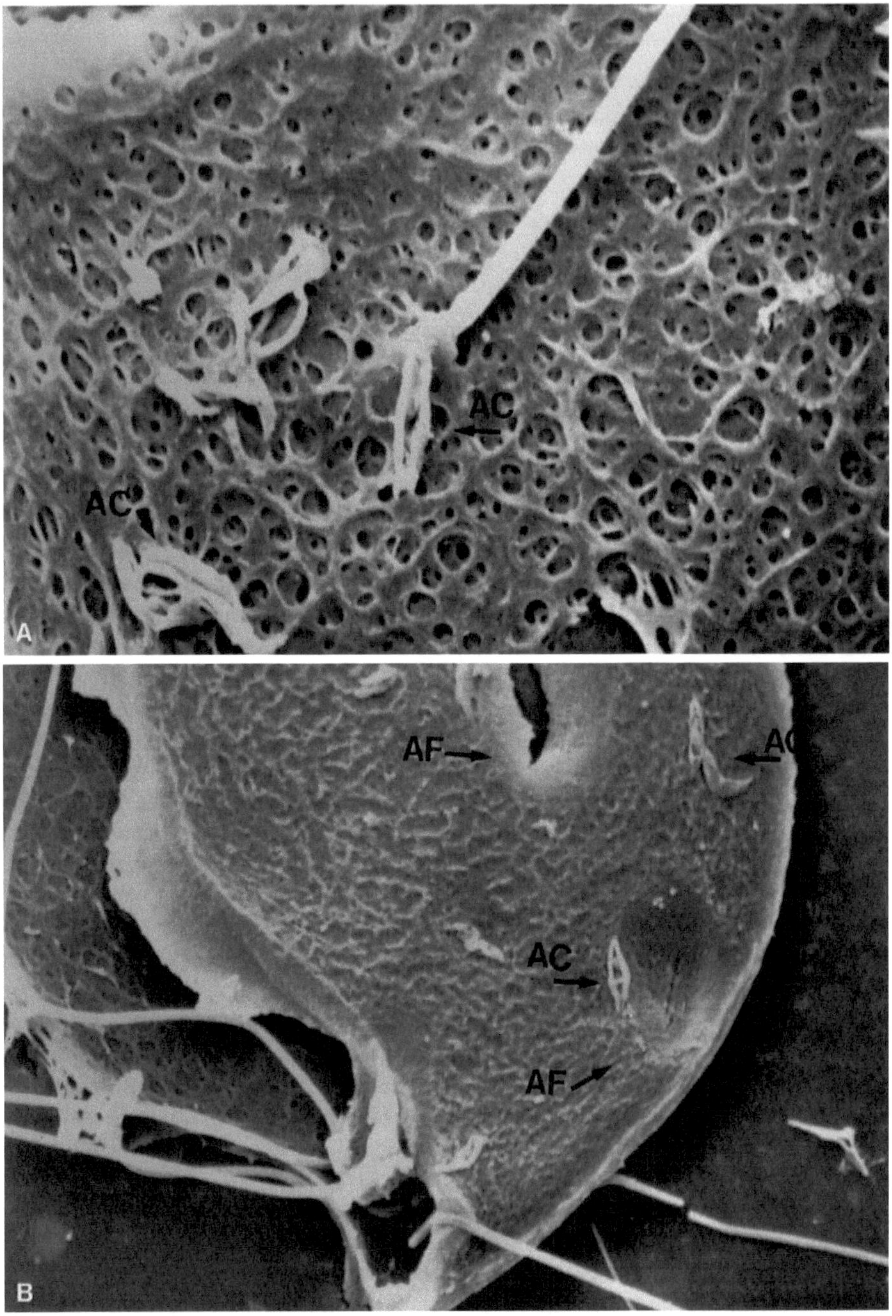

Fig. 4 A, B. Isolated hamster zonae were exposed to capacitated spermatozoa for 160 min, b which time sperm had undergone an acrosome reaction leaving detached acrosomal caps *(AC* bound to the zona surface. **A** A spermatozoon is shown entering the zona. **B** An inverted zon showing two acrosomal furrows *(AF)* where sperm have entered

ago by Dr. Hartmann and the present author, who found that binding occurred within 1 min to isolated zonae, but that 30-40 min were required for whole eggs (Hartmann et al. 1972; Hartmann and Hutchison 1974). Hartmann (1983; Hartmann et al 1982) developed an assay based on this delay in binding to show that several peptides are released, one from the vitellus, during the prebinding period of gamete association. These peptides which they hypothesized ensured the 30-40 min delay in gamete binding that was characteristic of whole eggs. These are original and fascinating studies that need to be extended.

However, it should be understood that a viable vitellus is not necessary for sperm entry. After prolonged incubation, as we have seen, hamster sperm will undergo an acrosome reaction and enter zonae in the absence of the vitellus. Salt-stored hamster eggs in which the vitellus is shrunken are nevertheless penetrated by capacitated hamster sperm (Yanagimachi et al. 1979), indicating that a viable vitellus is not needed to complete these processes. In the studies of Yanagimachi et al. (1979, 1981) the capacitated sperm were incubated with the salt-stored eggs for 5-7 h before the results were recorded, so that it is not known whether the binding that occurred was early or late.

We have seen that the vitellus is required to produce the apparent delay in binding (30-40 min) characteristic of whole eggs, and it is known that penetration into the vitellus begins to occur within 5 min thereafter (Hartmann and Hutchison 1976; Cherr et al. 1986). Hence, it is likely that what appears as delayed binding may actually be the first stage of penetration. As a working hypothesis we suggest that after the acrosome reaction the sperm may be held loosely to the egg by a tether that has been observed connecting it to the cap (Gwatkin et al. 1976) and in this state may be readily removed by pipeting. The acrosome caps, less subject to shear because a greater proportion of their surface is bound, would remain on the zona. With vitellus-free zonae the acrosome reaction would be delayed (see above), and the sperm would stay in place and appear as early binding.

Conclusions

It is apparent that considerable work needs to be done before we have a complete picture of the zona binding sites on the mammalian sperm. It is now well established that they are located on the plasma membrane of the sperm head. Following the acrosome reaction on the surface of the zona, binding may continue by the postacrosomal plasma membrane and by the tether connecting it to the acrosome cap. Binding may then occur by the acrosomal matrix as the sperm begins to enter the zona and to penetrate the egg. Binding does not occur to the inner acrosomal membrane. Most studies have ignored the acrosomal matrix because it is lost rapidly in the absence of protease inhibitors. When it is retained, labeling of the matrix may be demonstrated (as in Fig. 2). Tesarik et al. (1985) have reported that naturally occurring sperm isoantibodies in an infertile woman react specifically with the human sperm acrosomal matrix. In Fig. 1a of their publication the reacting matrix appears to extrude through the plasma membrane where it could be involved in binding even prior to the acrosome reaction. Further work must be carried out to define the binding molecules in both the plasma membrane and the acrosomal matrix. It is now possible to isolate the latter as an intact structure so that its specific binding components can be identified and their structures determined. This research is now underway in our laboratories (Hyatt and Gwatkin 1988).

References

Ahuja KK (1982) Fertilization studies in the hamster. Exp Cell Res 140: 353-362

Ball GD, Bellin ME, Ax RL, First NL (1982) Glycosaminoglycans in bovine cumulus-oocyte complexes: morphology and chemistry. Mol Cell Endocrinol 28: 113-122

Benau DA, Storey BT (1987) Characterization of the mouse sperm plasma membrane zona-binding site sensitive to trypsin inhibitors. Biol Reprod 36: 282-292

Bleil JD, Wassarman PM (1980) Mouse sperm-egg interaction: identification of a glycoprotein in mouse egg zonae pellucidae possessing receptor activity for sperm. Cell 20: 873-882

Bleil JD, Wassarman (1983) Sperm-egg interactions in the mouse: sequence of events and induction of the acrosome reaction by a zona pellucida glycoprotein. Dev Biol 95: 317-324

Bleil JD, Wassarman PM (1986) Autoradiographic visualization of the mouse egg's sperm receptor bound to sperm. J Cell Biol 102: 1363-1371

Brown CR, Cheng WKT (1986) Changes in composition of the porcine zona pellucida during development of the oocyte to the 2- to 4-cell embryo. J. Embryol. Exp. Morphol. 92: 183-191

Brown CR, Jones R (1987) Binding of zona pellucida proteins to a boar sperm polypeptide of M_r 53 000 and identification of zona moieties involved. Development 99: 333-339

Cherr GN, Lambert H, Meizel S, Katz DF (1986) In vitro studies of the golden hamster sperm acrosome reaction: completion on the zona pellucida and induction by homologous soluble zonae pellucidae. Dev Biol 114: 119-131

Cummins JM, Yanagimachi R (1986) Development of ability to penetrate the cumulus oophorus by hamster spermatozoa capacitated in vitro, in relation to the timing of the acrosome reaction. Gamete Res 15: 187-212

Dietl J, Czuppon A, Weichert K, Mettler L (1983) Identification of a sperm receptor-glycoprotein from the procine zona pellucida. Hoppe Seylers Z Physiol Chem 364: 261-267

Endo Y, Schultz RM, Kopf GS (1987) Effects of phorbol esters and a diacylglycerol on mouse eggs: inhibition of fertilization and modification of the zona pellucida. Dev Biol 119: 199-209

Florman HM, Wassarman PM (1985) O-linked oligosaccharides of mouse egg ZP3 account for it sperm receptor activity. Cell 41: 313-324

Florman HM, Bechtol KB, Wassarman PM (1984) Enzymatic dissection of the functions of the mouse egg's receptor for sperm Dev Biol 106: 243-255

Fowler RE, Kaufman MH, Grainge C (1986) The secretions of the cumulus-oocyte complex in relation to fertilization and early mouse embryonic development: a histochemical study. Histochem J 18: 541-550

Fox LL, Shivers CA (1975) Immunologic evidence for addition of oviductal components to the hamster zona pellucida. Fertil Steril 26: 599-608

Friess AE, Toepfer-Petersen E, Nguyen H, Schill W-B (1987a) Electron microscopic localization of a fucose-binding protein in acrosome reacted boar spermatozoa by the fucosyl-peroxidase-gold method. Histochemistry 86: 297-303

Friess AE, Toepfer-Petersen E, Schill W-B (1987b) Fracture labeling of boar spermatozoa for the fucose-binding-protein (FBP). Histochemistry 87: 181-183

Gao B, Klein LE, Britten RJ, Davidson EH (1986) Sequence of mRNA coding for bindin, a species-specific sea urchin sperm protein required for fertilization. Proc Natl Acad Sci USA 83: 8634-8638

Green DPL (1978) The activation of proteolysis in the acrosome reaction of guinea pig sperm. J Cell Sci 32: 153-164

Gwatkin RBL (1978) Cell-cell interactions in mammalian fertilization. Birth Defects 14: 363-376

Gwatkin RBL (1981) Sugar composition of the porcine zona pellucida. J Cell Biol 91: 170a

Gwatkin RBL, Anderson OF (1973) Effect of glycosidase inhibitors on the capacitation of hamster spermatozoa by cumulus cells in vitro. J Reprod Fertil 35: 565-567

Gwatkin RBL, Klein D (1985) Binding of solubilized zona by porcine sperm as they undergo the acrosome reaction. J Androl 6: 28-P

Gwatkin RBL, Williams DT (1977) Receptor activity of the hamster and mouse solubilized zona pellucida before and after the zona reaction. J Reprod Fertil 49: 55-59

Gwatkin RBL, Williams DT (1978) Bovine and hamster zona solutions exhibit receptor activity for capacitated but not for noncapacitated sperm. Gamete Res 1: 259-263

Gwatkin RBL, Carter HW, Patterson H (1976) Association of mammalian sperm with the cumulus cells and the zona pellucida studied by scanning electron microscopy. Scan Electron Micros 2: 379-384

Hartmann JF (1983) Mammalian fertilization: gamete surface interactions in vitro. In: Hartmann JF (ed) Mechanism and control of animal fertilization. Academic, New York, pp 325-364

Hartmann JF, Hutchison CF (1974) Nature of the prepenetration contact interactions between hamster gametes in vitro. J Reprod Fertil 36: 49-57

Hartmann JF, Hutchison CF (1976) Surface interactions between mammalian sperm and egg: variation of spermatozoa concentration as a probe for the study of binding in vitro. J Cell Physiol 88: 219-226

Hartmann JF, Gwatkin RBL, Hutchison CF (1972) Early contact interactions between mammalian gametes in vitro: evidence that the vitellus influences adherence between sperm and zona pellucida. Proc. Natl Acad Sci USA 69: 2767-2769

Hartmann JF, Hutchison CF, Vandlen RL (1982) Modulation of fertilization by the vitellus after interaction with peptides released by hamster sperm-zona pellucida contact. Dev Biol 93: 145-151

Henderson CJ, Hulme MJ, Aitken RJ (1987) Analysis of the biological properties of antibodies raised against intact and deglycosylated porcine zonae pellucidae. Gamete Res 16: 323-341

Huang TTF, Fleming AD, Yanagimachi R (1981) Only acrosome-reacted spermatozoa can bind to and penetrate zona pellucida: a study using the guinea pig. J Exp Zool 217: 287-290

Huang TTF, Ohzu E, Yanagimachi R (1982) Evidence suggesting that L-fucose is part of a recognition signal for sperm-zona pellucida attachment in mammals. Gamete Res 5: 355-361

Huneau D, Harrison RAP, Fléchon J-E (1984) Ultrastructural localization of proacrosin and acrosin in ram spermatozoa. Gamete Res 9: 425-440

Hyatt H, Gwatkin RBL (1988) Characterization of isolated acrosomal matrices from hamster spermatozoa. J Reprod Fert 83: 419-429

Jones R, Brown CR, Lancaster RT (1988) Carbohydrate-binding properties of boar sperm proacrosin and assessment of its role in sperm-egg recognition and adhesion during fertilization. Development 102: 781-792

Kuzan FB, Fleming AD, Seidel GE (1984) Successful fertilization in vitro of fresh intact oocytes by perivitelline (acrosome-reacted) spermatozoa of the rabbit. Fertil Steril 41: 766-770

Lee MA, Storey BT (1985) Evidence for plasma membrane impermeability to small ions in acrosome-intact mouse spermatozoa bound to mouse zonae pellucidae, using an aminoacridine fluorescent pH probe: time course of the zona-induced acrosome reaction monitored by both chlortetracycline and pH probe fluorescence. Biol Reprod 33: 235-246

Lenz RW, Bellin ME, Ax RL (1983) Rabbit spermatozoa undergo an acrosome reaction in the presence of glycosaminoglycans. Gamete Res 8: 11-19

Léveillé M-C, Roberts KD, Chevalier S, Chapdelaine A, Bleau G (1987) Uptake of an oviductal antigen by the hamster zona pellucida. Biol Reprod 36: 227-238

Meizel S, Turner KO (1985) Glycosaminoglycans stimulate the acrosome reaction of previously capacitated hamster sperm. J Cell Biol 99: 261a

Myles DG, Primakoff P (1984) Localized surface antigens of guinea pig sperm migrate to new regions prior to fertilization. J Cell Biol 99: 1634-1641

Naz RK, Phillips TM, Rosenblum BB (1986) Characterization of the fertilization antigen 1 for the development of a contraceptive vaccine. Proc Natl Acad Sci USA 83: 5713-5717

O'Rand MG, Fisher SJ (1987) Localization of zona pellucida binding sites on rabbit spermatozoa and induction of the acrosome reaction by solubilized zonae. Dev Biol 119: 551-559

O'Rand MG, Irons GP, Porter JP (1984) Monoclonal antibodies to rabbit sperm autoantigens. I. Inhibition of in vitro fertilization and localization on the egg. Biol. Reprod 30: 721-729

O'Rand MG, Matthews JE, Welch JE, Fisher SJ (1985) Identification of zona binding proteins of rabbit, pig, human, and mouse spermatozoa on nitrocellulose blots. J Exp Zool 235: 423-428

Peterson RN, Russell LD, Bundman D, Freund M (1980) Binding of boar spermatozoa to porcine oocytes: evidence for specific membrane receptors for the zona pellucida. Science 84: 144-156

Peterson RN, Henry L, Hunt W, Saxena N, Russell LD (1985) Further characterization of boar sperm plasma membrane proteins with affinity for the porcine zona pellucida. Gamete Res 12: 91-100

Peterson RN, Hunt WP, Henry LH (1986) Interaction of boar spermatozoa with porcine oocytes: increase in proteins with high affinity for the zona pellucida during epididymal transit. Gamete Res 14: 57-64

Phillips DM, Shalgi R. (1982) Sperm penetration into rat ova fertilized in vivo. J Exp Zool 221: 373-378

Primakoff P, Hyatt H, Myles DG (1985) A role for the migrating sperm surface antigen PH-20 in guinea pig sperm binding to the egg zona pellucida. J Cell Biol 101: 2239-2244

Ringuette MJ, Sobieski DA, Chamow SM, Dean J (1986) Oocyte-specific gene expression: molecular characterization of a cDNA coding for ZP3, the sperm receptor of the mouse zona pellucida. Dev Biol 83: 4341-4345

Ruiz-Bravo N, Earles D, Lennarz WJ (1986) Identification and partial characterization of sperm receptor associated with the newly formed fertilization envelope from sea urchin eggs. Dev Biol 117: 204-208

Sacco AG, Subramanian MG, Yurewicz EC (1984) Association of sperm receptor activity with a purified pig zona antigen. J Reprod Immunol 6: 89-103

Sacco AG, Yurewicz EC, Subramanian MG (1986) Carbohydrate influences the immunogenic and antigenic characteristics of the ZP3 macromolecule (M_r 55000) of the pig zona pellucia. J Reprod Fertil 76: 575-586

Saling PM (1981) Involvement of trypsin-like activity in binding of mouse spermatozoa to zonae pellucidae. Proc Natl Acad Sci USA 78: 6231-6235

Saling PM (1986) Mouse sperm antigens that participate in fertilization. IV. A monoclonal antibody prevents zona penetration by inhibition of the acrosome reaction. Dev Biol 117: 511-519

Saling PM, Storey BT (1979) Mouse gamete interactions during fertilization in vitro. Chlortetracycline as a fluorescent probe for the mouse sperm acrosome reaction. J Cell Biol 83: 544-555

Shalgi R, Matityahu A, Nebel L (1986) The role of carbohydrates in sperm-egg interaction in rats. Biol Reprod 34: 446-452

Shur BD, Hall NG (1982a) Sperm surface galactosyltransferase activities during in vitro capacitation. J Cell Biol 95: 567-573

Shur BD, Hall NG (1982b) A role for mouse sperm surface galactosyltransferase in sperm binding to the egg zona pellucida. J Cell Biol 95: 574-579

Storey BT, Lee MA, Muller C, Ward CR, Wirtshafter DG (1984) Binding of mouse spermatozoa to the zonae pellucida of mouse eggs in cumulus: evidence that the acrosomes remain substantially intact. Biol Reprod 31: 1119-1128

Sullivan R, Bleau G (1985) Interaction of isolated components from mammalian sperm and egg. Gamete Res 12: 101-116

Talbot P. (1984) Hyaluronidase dissolves a component in the hamster zona pellucida. J Exp Zool 229: 309-316

Tesarik J, Travnik P, Mecova L, Pilka L (1985) Binding of antisperm isoantibodies from a woman at risk for immunologic infertility to intraacrosomal components of donor spermatozoa. Fertil Steril 44: 138-141

Toepfer-Petersen E, Friess AE, Nguyen H, Schill W-B (1985) Evidence for a fucose-binding protein in boar spermatozoa. Histochemistry 83: 139-145

Töpfer-Petersen E, Henschen A (1988) Zona pellucida-binding and fucose-binding of boar sperm acrosin is not correlated with proteolytic activity. Biol Chem Hoppe Seyler 369: 69-76

Vacquier VD, Moy GW (1977) Isolation of bindin: the protein responsible for adhesion of sperm to sea urchin eggs. Proc Natl Acad Sci USA 74: 2456-2460

Yanagimachi R, Lopata A, Odom CB, Bronson RA, Mahi CA, Nicolson GL (1979) Retention of biological characteristics of zona pellucida in highly concentrated salt solution: the use of salt stored eggs for assessing the fertilizing capacity of spermatozoa. Fertil Steril 31: 562-574

Yanagimachi R, Okada A, Tung KSK (1981) Sperm autoantigens and fertilizations. II. Effects of anti-guinea pig sperm autoantibodies on sperm-ovum interactions. Biol. Reprod. 24: 512-518

Fractionation and Characterization of the Glycoproteins of Zona Pellucida

M. Nakano

Introduction

The zona pellucida that surrounds the mammalian oocyte plays a physiological role in fertilization and in development before implantation (Wolf 1982; Dunbar 1983). This extracellular coat is thought to be the region in which polyspermy is blocked, species specificity in sperm-egg interactions is determined, and the embryo is physically protected until implantation. However, the molecular mechanisms of these biological functions are not yet fully understood. The zona pellucida has also been receiving interest regarding the biosynthesis of glycoproteins and their transport through biomembrane (Wassarman et al. 1981, 1985; Wassarman 1983).

Since the zona is composed of various kinds of glycoproteins, it is indispensable to purify the individual molecular species in order to ascertain the relationships between structures and functions. Establishing methods for fractionation of the glycoproteins and for isolation of pure zonae pellucidae on a large scale is therefore very important. Biological functions of the glycoproteins in mouse zona have been well studied, and partial structural analyses of porcine zona proteins have also been performed. This chapter deals mainly with the zonae of these two species.

Isolation of the Zonae Pellucidae

Mouse Zonae Pellucidae

Rodent gametes are suitable for the study of mammalian fertilization in vitro. Ovarian eggs in various stages of growth (Bleil and Wassarman 1980c), ovulated eggs, and embryos can be obtained from ovary or oviduct (Schultz et al. 1979a, b; Aitken and Richardson 1981). However, a method for large-scale isolation of mouse, rat, and hamster zonae has not been developed, since rodent zonae are smaller in width and more fragile than are those of domestic animals (Dunbar 1983).

In preparation of mouse zonae from ovarian eggs, zona-encased oocytes are released into medium by puncturing the ovarian follicles of the excised ovaries with fine needles. By this procedure fully grown eggs (80–85 μm in diameter) and growing eggs (40–50 μm) are obtained from adult (8- to 12-week-old) and juvenile (6-day-old) mice, respectively (Bleil and Wassarman 1980a; Greve et al. 1982).

Ovulated eggs, arrested at metaphase II, are obtained from oviducts of females which have been superovulated by injection of pregnant mare serum (PMS; 5-10 IU), followed by the same dose of human chorionic gonadotropin (HCG) 48 h later. The mice are killed at 14-16 h after HCG injection, and eggs are recovered by removal of the oviduct. To obtain embryos the mice are also superovulated. After HCG injection the females are caged overnight with the males, and matings are detected by the presence of vaginal plug on the morning. This is designated as day 0 of pregnancy. Two-cell and eight-cell embryos are collected by flushing oviducts with culture medium on day 1 and day 3, respectively.

These eggs or embryos dispersed in medium are washed several times by transference into fresh medium using mouth-operated glass pipet under a stereoscopic microscope. If necessary, adherent cumulus cells are dispersed by incubation with hyaluronidase (300 U/ml medium) before washing (Shimizu et al. 1982). Eggs or embryos are then sucked into a fine glass pipet having a bore size less than the diameter of the eggs, and the zonae freed from oocyte contents are again washed several times with medium. By this procedure about 50 zonae are isolated per hour of work.

Mouse zonae are also obtained from ovulated eggs and two-cell embryos by dissolution of oocyte contents with detergent (Shimizu et al. 1982, 1983; Shimizu and Ito 1986). After treatment with hyaluronidase and washing with fresh medium, eggs or embryos are suspended in Hanks's medium or modified Brinster's medium containing NP40 (0.2%), bovine serum albumin (BSA; 0.01%-0.02%), ethylene diaminetetraacetate (EDTA; 10 m*M*), phenylmethanesulfonic acid fluoride PMSF (0.1-1 m*M*), and lima bean trypsin inhibitor (0.001%). After three cycles of freezing and thawing the zonae with no apparent cellular debris are collected by micropipet and washed thoroughly with the medium without detergent and inhibitors. The recovery rate of zonae by this method from ovulated eggs is 30%-40%, while that from two-cell embryo is more than 50%.

Relatively large numbers of mouse zonae from ovarian eggs are isolated by sucrose (Brauer and Sotelo 1983) or Percoll (Bleil and Wassarman 1986) gradient centrifugation of ovarian homogenates. In the Percoll method the homogenate in a triethanolamine buffer (pH 8.5) containing DNase and hyaluronidase is brought to 1% NP40 and 0.1 m*M* PMSF. After addition of deoxycholate to make 1%, the homogenate is mixed with a buffer containing Percoll and centrifuged in 50% Percoll solution at 25 000 rpm for 45 min at 4 °C. The band of zonae (density of about 1.02 g/ml) separated from those of ovarian tissue (1.10-1.14 g/ml) is recovered by puncturing the side of the centrifuge tube with a syringe needle and is washed twice with the buffer by centrifugation. By this procedure $5\text{-}20 \times 10^3$ zonae are obtained from 20-30 mice. The zona of a fully grown egg has a width of 5-7 μm, and its dry weight and protein content are 6.0 ng and 3.0-4.8 ng per zona, respectively (Loewenstein and Cohen 1964; Dunbar 1983; Wassarman et al. 1985).

Since a large amount of protein of mouse zona is difficult to obtain, isolated zonae are often radiolabeled with the reagents for iodination, which allows the preparation of proteins at high specific activity. For labeling, diazonium salt of [^{125}I]iodosulfanilic acid (Bleil and Wassarman 1980 a, c), ^{125}I-labeled Bolton-Hunter reagent (*N*-hydroxysucciniimide ester of iodinated *p*-hydroxyphenylpropionic acid; Greve et al. 1982; Florman et al. 1984; Florman and Wassarman 1985), or

$Na^{125}I$ with chloramine T (Bleil and Wassarman 1986) has been used; average specific activity of the modified product with three reagents is about 1.6×10^3, 2.0×10^4, and 4.8×10^5 cpm per zona pellucida, respectively.

Before submission to fractionation methods or immunological characterization radiolabeled or unlabeled zonae are solubilized under the following various conditions: (a) freezing (−70 °C) and thawing (60 °C) four times in Tris-HCl, pH 8.0 or pH 6.8, containing sodium dodecyl sulfate (SDS; Bleil and Wassarman 1980a; Greve et al. 1982); (b) incubation in phosphate-buffered solution (PBS) containing 0.2% polyvinylpyrrolidone-40 adjusted to pH 3.0 with HCl (Bleil et al. 1981); (c) incubation in ammonium acetate buffer, pH 2.5 or 5.0, at 37 °C for 30 min (Greve and Wassarman 1985); (d) incubation in 8 *M* urea in PBS at 37 °C (Aitken et al. 1982); (e) heating at 68 °C in Dulbecco's PBS, pH 7.2 (Aitken et al. 1982); (f) incubation in "solubilization buffer" for SDS-PAGE (polycrylamide gel electrophoresis) in the presence or absence of dithiothreitol (DTT; Bleil and Wassarman 1980a; Greve and Wassarman 1985).

Alternatively, mouse zonae can be selectively solubilized from eggs with Tyrode's acidic buffer at pH 2.5 (Nicolson et al. 1975) or 5-m*M* NaH_2PO_4 at pH 2.5 (Bleil and Wassarman 1980c; Roller and Wassarman 1983; Florman et al. 1984). In some cases, radiolabeling of zona proteins with ^{125}I-iodinated or tritium-labeled reagents ([^3H]dansylchloride and [^3H]dinitrofluorobenzene) is also performed after solubilization of the zona (Bleil and Wassarman 1980a).

Porcine Zonae Pellucidae

Ovulated eggs and fertilized embryos are difficult to obtain on a large scale from domestic animals. However, many zonae which are relatively large and rigid to rodent zonae can be mechanically isolated from their ovarian eggs. Several research groups have developed methods for large-scale isolation of porcine zonae, since ovaries are easily obtainable from slaughtered gilts. Consequently, 1000–2000 ovaries can be processed at a time (Noda et al. 1981; Yurewicz et al. 1987); $0.5\text{–}6 \times 10^4$ zonae per 100 ovaries are obtained, and each zona is about 16 μm in width and contains about 33 ng of protein (Dunbar et al. 1980; Dunbar 1983).

To rupture the ovarian follicles a meat grinder (Oikawa 1978; Tsunoda and Sugie 1979; Yurewicz et al. 1987), multineedled flower base (Dunbar et al. 1980), or series of ganged razor blades (Dunbar and Raynor 1980; Dunbar et al. 1980; Gwatkin et al. 1980; Henderson et al. 1987) has been used. Although thawed ovaries are usually submitted for rupture, slicing of frozen ovaries with a razor blade is an efficient means for reducing contamination of cumulus debris in the released follicular contents (Nakano et al. 1987b). These contents containing eggs are suspended in PBS and filtered through a series of stainless meshes (Tsunoda and Sugie 1979; Nakano et al. 1987b), nylon mesh (Dunbar and Raynor 1980; Dunbar et al. 1980, 1981; Dietl et al. 1983c) or Saran mesh screens (Noda et al. 1981). After the last filtration through a mesh in the range of 63–105 μm the trapped eggs, having a diameter of about 157 μm and small clumps of follicle cells, are washed out with PBS; these washing procedures are repeated two to five times. Sweeping with glass wool (Oikawa 1978; Noda et al. 1981; Nakano et al. 1987b) or human hair (Tsunoda and Sugie 1979) or digestion with collagenase (Gwatkin et al. 1980; Dietl et al. 1983c) is also performed to remove the debris of follicle cells. Furthermore, centrifugation in sucrose or Percoll solution has been reported to be efficient for this purpose. Noda et al. (1981) overlayered the suspension of crude eggs in PBS on 50% sucrose solution and centrifuged at 1600 *g* for 30 min. Pure eggs are retained in the interphase between PBS and sucrose layers. Recently Henderson et al. (1987) purified eggs by applying a discontinuous Percoll density layer (40%-20%-10%). In the centrifugation of this layer at 2000 *g* for 20 min at 4 °C the eggs remain at the top of the gradient. The purified eggs obtained by these various methods are thoroughly washed with PBS.

The zonae are then removed from oocyte contents by a gentle homogenization with a Potter-Elvehjem homogenizer (Dunbar and Raynor 1980; Dunbar et al. 1980; Gwatkin et al. 1980; Noda et al. 1981; Hedrick and Wardrip 1986a; Henderson et al. 1987) or by sucking into a fine glass pipet having a bore size less than the diameter of the egg (Tsunoda and Sugie 1979; Nakano et al. 1987b; Hedrick et al. 1987). The zonae thus freed from oocytes are trapped on the mesh of 41-63 μm (Gwatkin et al. 1980; Dunbar and Raynor 1980; Hedrick and Wardrip 1986a; Nakano et al. 1987b) and washed extensively with PBS. Prior to fractionation the zonae are solubilized by following various heating procedures: (a) in distilled water or PBS at 78 °C-80 °C for 20-35 min (Gwatkin et al. 1980; Hedrick and Wardrip 1986b; Henderson et al. 1987); (b) in 5-m*M* ammonium bicarbonate, pH 8.2, at 70 °C for 30 min (Hedrick and Wardrip 1986a); (c) in 20-m*M* Tris buffer, pH 7.0, at 73 °C for 20 min, in sample buffer for SDS-PAGE at 95 °C for 5 min (Dunbar et al. 1981).

Brown and Cheng (1986) and Hedrick et al. (1987) recovered fertilized embryos as well as ovulated unfertilized eggs from porcine oviducts and isolated zonae pellucidae. Since only a small number of zonae is obtainable, the zonae are radiolabeled with ^{125}I-labeled Bolton-Hunter reagent before solubilization (Brown and Cheng 1986) or with Na^{125}I using IODO-GEN (1,3,4,6-tetrachloro-3α, 6α-diphenylglycoluril) after solubilization (Hedrick et al. 1987).

In 1980 Dunbar et al. presented the results of detailed chemical analyses of isolated zonae from ovarian eggs. They determined amino acid, carbohydrate, and free fatty acid composition together with sulfate and phosphate contents. The amino acid composition was similar to that of unfractionated mouse zona proteins.

Zonae Pellucidae of Other Mammals

The egg diameter and the zona width differ among various mammals (Dunbar 1983). For example, the average diameter (±SD) of hamster and bovine eggs is 118.6±1.3 μm and 146.2±5.7 μm, respectively (Gwatkin et al. 1980). However the isolation procedures for zonae are almost identical. Zonae pellucidae of rat (Sacco et al. 1981) and golden hamster (Hartmann and Gwatkin 1971; Phillips and Shalgi 1980; Hartmann et al. 1982; Sullivan and Bleau 1985) have been isolated from ovarian eggs or superovulated eggs in oviducts by essentially the same mechanical procedure (using a fine glass pipet) as in the case for mouse zonae.

The zonae of rabbit (Dunbar et al. 1981), bovine (Gwatkin et al. 1980), squirrel monkey (Sacco et al. 1981), and human (Sacco 1977) are also prepared from ovaries by the mechanical method. About 8×10^3 zonae and 1.2×10^3 zonae are obtained from 100 ovaries of rabbit and bovine, respectively.

Fractionation of Zona Proteins

Mouse Zona Proteins

Electrophoresis. In 1980 Bleil and Wassarman reported a method for fractionation of mouse zona glycoproteins (Bleil and Wassarman 1980a). They separated zona proteins labeled with [^{125}I]iodosulfanilic acid in nonreduced state into three components by one- and two-dimensional PAGE. One-dimensional SDS-PAGE using 7% or 9% acrylamide in separating gel (Laemmli 1970) reveals that mouse zona from a fully grown egg is composed of only three kinds of glycoprotein (ZP1, ZP2, and ZP3). Although the bands in acrylamide gel are broad, reflecting heterogeneous glycosylation of the proteins, their apparent molecular weights are estimated to be 200 kd for ZP1, 120 kd for ZP2, and 83 kd for ZP3 under nonreducing condition. These three glycoproteins account for virtually all the mass of the zona pellucida, and ZP1, ZP2, and ZP3 represent about 36%, 47%, and 17%, respectively, of the protein in the zona pellucida (Bleil and Wassarman 1980a).

Diagonal electrophoresis, performing "reduction electrophoresis" in the second dimension, reveals that ZP1 migrates as a single spot, with an apparent molecular weight of 130 kd under reducing condition. This fact suggests that ZP1 exists as a dimer forming interchain disulfide linkage in nonreduced state. It is also shown that the apparent molecular weight of ZP3 in reduced state changes from 83 to 85 kd, and that that of ZP2 does not change, indicating the absence of interchain disulfide linkage in these two components (Bleil and Wassarman 1980a). By SDS-PAGE the zonae of growing oocytes from 8- to 15-day-old mice are shown to contain a set of three components (Bleil and Wassarman 1980b), while an another band (ZPf) between ZP1 and ZP2 is detected in the gel of the zonae from two-cell embryos in addition to the original three bands (Shimizu and Ito 1986).

Two-dimensional PAGE of O'Farrel's system (O'Farrell 1975) shows that each of the zona proteins modified by iodosulfanilic acid - ^{125}I-labeled ZP1, ZP2, and ZP3 - also presents a single spot in the second dimension of isoelectric focusing (IEF) gel, forming a linear pH gradient from 3.5 to 8.0, and that their isoelectric points are 5.9, 4.3, and 3.7, respectively (Bleil and Wassarman 1980a). Although these values are not necessarily identical to the pI values of the unmodified proteins, the acidic nature is thought to be in part due to the presence of sialic acid.

After electroelution of the gel of SDS-PAGE containing each zona protein, amino acid analyses were performed (Bleil and Wassarman 1980a); the results are shown in Table 1. It is noticeable that cystine, which is thought to participate in the maintenance of the zona structure (Yanagimachi 1977), is abundant in each component, particularly in ZP3, and that ZP1 has an extremely high glycine content amounting to 21 mol%. Tyrosine and aspartic acid are abundant in ZP1 and

Table 1. Amino acid composition (mol%) of three components of mouse zona protein

	ZP1	ZP2	ZP3
Asx	6.3	9.9	6.2
Thr	2.8	5.5	4.0
Ser	7.2	9.1	6.9
Glx	10.6	10.4	10.7
Pro	5.4	7.2	8.3
Gly	20.9	6.7	13.6
Ala	4.7	6.3	5.0
½ Cys	4.5	4.1	7.1
Val	8.9	8.2	9.2
Met	1.7	1.2	1.4
Ile	4.2	5.2	4.4
Leu	7.0	9.1	8.3
Tyr	3.2	0.8	0.8
Phe	6.2	5.7	5.8
Lys	2.7	4.4	3.5
His	1.2	2.2	0.8
Arg	2.9	3.9	4.0

(From Bleil and Wassarman 1980a)
Tryptophan was not determined.

ZP2, respectively, while ZP2 has a low content of glycine. These results together with those of electrophoresis suggest that each of the three zona proteins represents a unique polypeptide chain.

Column Chromatography. Good resolution of mouse zona proteins (MZP) has also been achieved by high performance liquid chromatography (HPLC) under the same condition used for fractionation of porcine zona proteins (PZP) (Nakano et al. 1987b). When MZP in nonreduced state are applied to a Toyo Soda (TSK) gel G3000SW column, three peaks (MZP1-3) are detected by fluorescence monitor as well as by UV monitor (Fig. 1), indicating the presence of tryptophan residues in the three components (Nakano et al., unpublished result). The molecular weight of these components is estimated to be 142 kd for MZP1, 94 kd for MZP2, and 54 kd for MZP3, by comparison with the retention time of the molecular weight standards on the same column. Although these values are smaller than those estimated by electrophoresis, MZP1, 2, and 3 probably correspond to ZP1, 2, and 3 in electrophoresis, respectively. The discrepancy in apparent molecular weights may be due to a different property in the interaction between zona proteins and HPLC resin or acrylamide gel.

On the other hand, Bleil and Wassarman (1986) purified ^{125}I-labeled ZP3 from heat-solubilized mouse zonae by immunoadsorption to anti-ZP3 Sepharose, which is prepared by cross-linking BrCN-Sepharose to a monoclonal antibody directed against ZP3. Then ^{125}I-labelled ZP2 is purified from the unadsorped fraction to anti-ZP3 Sepharose by applying anti-ZP2 Sepharose prepared with a monoclonal antibody directed against ZP2.

Although the fractionation method for mouse glycoproteins has been established as described above, their biochemical analysis is difficult owing to the defi-

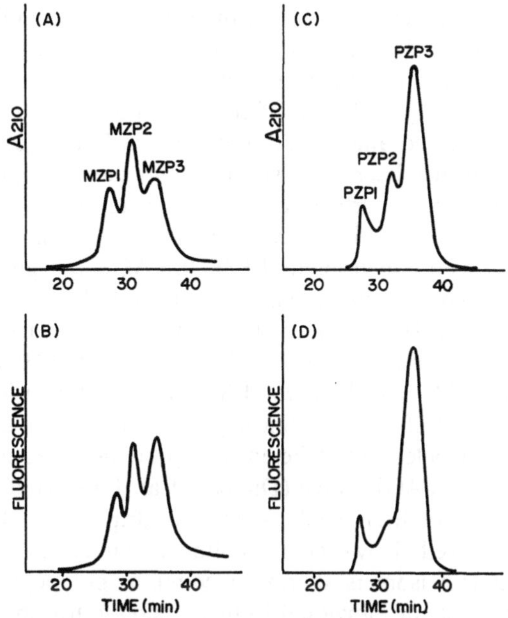

Fig. 1 A–D. Chromatographic separation of mouse and porcine zona proteins by HPLC under nonreducing conditions. Glycoproteins of heat-solubilized mouse zonae pellucidae (*MZP;* **A, B**) and porcine zonae pellucidae (*PZP;* **C, D**) were applied to a column of TSK gel G3000SW (7.5 mm inner diameter × 60 cm). The elution was performed with 3.5 m*M* SDS, 100 m*M* sodium sulfate, 50 m*M* sodium phosphate, pH 7.0, at a flow rate of 0.4 ml/min. The absorbance of the effluent at 210 nm **(A, C)** and the fluorescence of tryptophan residues with excitation at 278 nm and emission at 348 nm **(B, D)** were monitored successively

ciency in their amounts. Accordingly, not even the carbohydrate composition of each component, which gives some structural information on carbohydrate chain, has been determined. The presence of carbohydrates in each MZP from ovulated egg and two-cell embryo has only been ascertained by the analysis of reactivity with peroxidase-conjugated lectin or biotinyl lectin (Shimizu et al. 1982; Shimizu and Ito 1986).

Porcine Zona Proteins

Electrophoresis. SDS-PAGE of PZP was carried out at first by several research groups under reducing condition in 5.5%–7.5% acrylamide gel. However, the resolution is not satisfactory and a diffuse complex band having an apparent molecular weight ranging from 40 kd to 80 kd is detected (Dunbar and Raynor 1980; Dunbar et al. 1980, 1981; Gwatkin et al. 1980; Noda et al. 1981). When SDS-PAGE is performed under nonreducing condition, two major bands, one at 87 kd and another at 58 kd, are detected (Yurewicz et al. 1983). A smearing toward the top of the gel is also shown, suggesting the presence of at least two major and multiple minor components in porcine zona. By the gel system for analysis of membraneous glycoproteins of erythrocyte using 5.6% acrylamide gel (Fairbanks et al.

1971) a better resolution than that by Laemmli's system has been achieved (Nakano et al., unpublished observation). Noda et al. (1983) also separated the zona proteins in nonreduced state into two fractions (PZ-α, 60 kd; PZ-β, 96 kd) by preparative SDS-PAGE after partial purification by a Sephadex G-100 column and performed amino acid and carbohydrate analyses. No clear distinction between them has been observed either in their compositions or in their immunological properties. Furthermore, the N termini of the components are the same, and only glutamic acid is detected by dansyl method.

Two-dimensional IEF/SDS-PAGE has given interesting information on PZP (Sacco et al. 1981; Dunbar et al. 1981; Yurewicz et al. 1983; Hedrick and Wardrip 1986a). IEF in the first dimension in polyacrylamide gel, containing ampholine with a wide pH range of 3.0-10.0 or 3.5-10.0, is conducted at 300 V or 400 V for 16-22 h. For the second SDS-PAGE dimension electrophoresis is performed in the gel of 7% or a linear gradient (5%-15% or 10%-20%) of acrylamide. When ^{125}I-labeled or unlabeled zona proteins in reduced state are applied in two-dimensional PAGE, three (Dunbar et al. 1981; Hedrick and Wardrip 1981; Drell et al. 1984) or four (Hedrick and Wardrip 1986a, 1987) distinct bands are effectively resolved. However, it is shown that each component consists of about 10-25 charge isomers sharing a common polypeptide chain; the apparent molecular weights of individual isomers increase toward the acidic end of the gel, presumably owing to an alteration in the binding of SDS (Dunbar et al. 1981). This two-dimensional PAGE pattern of the protein isolated from a single egg is indistinguishable from that obtained en masse (Hedrick et al. 1987). Although this heterogeneity in charge is suggested to be due to the variation of the contents of sialic acid and sulfate (Hedrick and Wardrip 1981) or of sulfated and/or phosphorylated lactosaminoglycans attached to proteins (Yurewicz et al. 1987), its chemical basis is not fully understood. When the zona proteins in nonreduced state are applied to two-dimensional PAGE, two families comprising many charge isomers are also resolved (Yurewicz et al. 1983; Hedrick and Wardrip 1986a). The pI ranges of the iodinated zona proteins are summarized in Table 2.

Column Chromatography. Since porcine zonae can be obtained in relatively large quantities, column chromatography for fractionation of the components has been tried. The carboxymethylated zona proteins emerge as two peaks from a column of Sephacryl S-200, although the separation is not satisfactory (Noda et al. 1981). The predominant component, 55-kd protein, has also been isolated effectively by column chromatography. Yurewicz et al. (1983) prepared a purified pig zona antigen (PPZA) from heat-solubilized zonae by applying it successively to ultrogel AcA34 and diethylaminoethanol (DEAE) Bio-Gel A column in Tris buffer, pH 8.0. After treatment with SDS at 60 °C for 20 min this PPZA is further fractionated by a column of Sepharose CL-4B. When the column is developed with Tris buffer containing 0.1% SDS, three fractions are obtained, and SDS-PAGE of nonreduced specimens demonstrates that the second peak corresponds to 55 kd protein. They also isolated 55 kd protein by chromatography on the columns of Sephacryl S-400 and hydroxyapatite equilibrated with phosphate buffer containing 0.1% SDS (Sacco et al. 1986; Yurewicz et al. 1987).

Dietl et al. (1983a, c) reported the fractionation of zona proteins by HPLC for the first time. They collected the zonae by sieving techniques with nylon mesh after treatment with collagenase to remove cell debris (Gwatkin et al. 1980) and solubilized them with 0.3-*M* lithium-3,5-diiodosalicylate (LIS) in 0.05-*M* Tris-HCl, pH 7.5. The extracts with LIS were then mixed with phenol-water (1:1 by volume) to eliminate contamination, and the aqueous phase was lyophilized after dialysis against water. They applied this specimen successively to the two styro-gel protein columns of ^{60}I and ^{125}I. When 0.2-*M* formic acid in 6-*M* urea was used for mobile phase at a flow rate of 2.5 ml/min, a total of eleven fractions with apparent molecular weight between >100 kd and 8.2 kd were obtained within 15 min. They presumed that at least nine (fractions 1-9) of these are individual components (Dietl et al. 1983a). Furthermore, amino acid and N-terminal analyses as well as partial carbohydrate analysis of five fractions have been performed (Czuppon and Dietl 1985). Zona proteins were solubilized in LIS and separated into three fractions using immunoaffinity chromatography on immobilized anti-zona IgG (Dietl et al. 1983b). By SDS-PAGE the second fraction ZPII is found to contain two antigens, and the first and third fractions are indicated to be homogeneous. These four antigens probably correspond to the fractions 2, 3, 4, and 8 on HPLC, and two of them are supposed to express on the surface in intact zona. Although these procedures are unique, and the resolutional ability of this HPLC system is remarkable, there is a possibility of the presence of degraded products in the separated fractions by collagenase (Yurewicz et al. 1983; Hedrick and Wardrip 1986b). Since commercial collagenase contains many proteins, including trypsinlike protease, the zonae prepared by the use of collagenase may have protein molecules cleaved by this contaminated protease (Hedrick and Wardrip 1986b). Incidentally, treatment with acrosin or trypsin induces limited and specific proteolysis of the zona proteins without morphological change of the zona (Brown and Cheng 1985; Dunbar et al. 1985). Particular caution should be given to prevent proteolysis of the zona proteins in applying the enzyme method to purify the zona.

The proteins from mechanically isolated zonae are effectively fractionated into three families (PZP1-3) by HPLC on a gel filtration column (Nakano et al. 1987b). Chromatography is performed on a column of TSK gel G3000SW, and 0.1% SDS in PBS, pH 7.0, is used for the mobile phase at a flow rate of 0.2 ml/min. Even at a faster flow rate of 0.4 ml/min good resolution is achieved (Fig. 1), and each fraction emerges as a single peak on rechromatography in the same condition. The apparent molecular weights of the families are estimated to be 164 kd for PZP1, 99 kd for PZP2, and 55 kd for PZP3; these are present in an approximate weight ratio of protein moiety of 2:3:18, respectively. The recovery rate of total protein in this rapid and simple method is more than 90%, whereas only 45% of protein is recovered in preparative two-dimensional PAGE (Hedrick and Wardrip 1986a). When the PZP1 fraction of HPLC is applied to SDS-PAGE of Fairbanks' system (Fairbanks et al. 1971), a diffuse weak band is detected, and its apparent molecular weight is estimated to be in the range of 160-220 kd. This fact together with the differences in carbohydrate composition suggests that PZP1 is not comprised of aggregated products of other fractions. In HPLC minor components of high molecular weight probably coelute in the PZP1 fraction as a single peak. Hedrick and Wardrip indicated that minor high molecular weight components (>110 kd) and actin (45 kd) in the isolated zonae originate from the con-

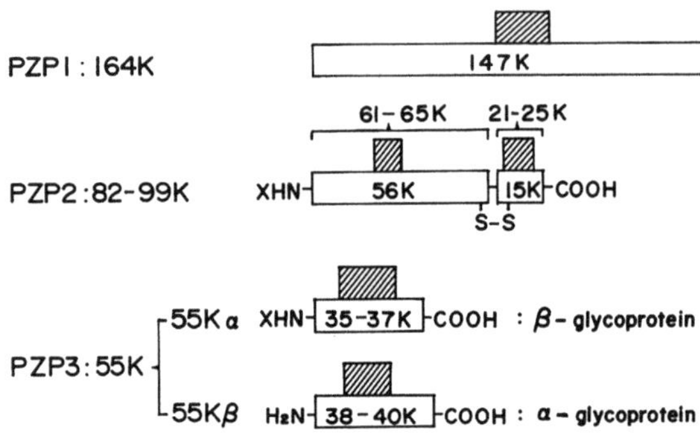

Fig. 2. Glycoproteins of porcine zona pellucida *(PZP)* (Hedrick and Wardrip 1987; Yurewicz et al. 1987; Nakano et al 1987b). The length of *open bars* and *crossed bars* represents the relative molecular weights of protein and carbohydrate moieties, respectively. PZP1 is a mixture of high molecular weight components. PZP2 has a blocked N terminus; two peptides are released from PZP2 in reducing condition by proteolysis. PZP3 is composed of two distinct glycoproteins. *K*, Kilodaltons; 55Kα having a blocked N-terminus is designated as β-glycoprotein by Yurewicz et al. (1987) while 55Kβ as α-glycoprotein

taminated villous processes of cumulus cells and oocytes, and that treatment of the zonae with 1% Triton X-100 is effective to remove these contaminations (Hedrick and Wardrip 1987). However, there is a fear that treatment with the detergents may affect the zona macromolecules. It has also been shown that the carbohydrate content in the first half of the PZP3 fraction in HPLC (Fig. 1C) is higher than that in the latter half (Nakano et al., unpublished observation). When the molecular weight of protein moiety is assumed to be 38-kd (see below), the contents of *N*-acetylglucosamine and galactose in the first half are 44 residues and 22 residues, respectively, while those in the latter half are 36 residues an 16 residues, respectively. A longer chain of lactosaminoglycan may be present in the first half. In contrast, the content of *N*-acetylgalactosamine in the first half (3 residues) is less than that in the latter (4 residues), reflecting the difference in the number of *O*-glycoside linkages to protein.

Molecular Weight. The apparent molecular weights of zona proteins in nonreduced and reduced states estimated by various research groups differ from each other (Menino and Wright 1979; Yurewicz et al. 1983; Hedrick and Wardrip 1981, 1986a; Henderson et al. 1987; Hedrick et al. 1987). The results given recently by three groups are summarized in Fig. 2. Under reducing condition the 65-kd and 25-kd components, which migrate at the dye front in SDS-PAGE, are derived from the 90-kd component, when proteolysis also occurs (Hedrick and Wardrip 1987). Furthermore, N termini of 90-kd and 65-kd components are suggested to be blocked. Accordingly, the 65-kd component has been inferred to originate from the N-terminal region of the 90-kd component the and 25-kd component from the C-terminal region.

On the other hand, when the 55-kd fraction purified by chromatography on Sephacryl S-400 and hydroxyapatite is treated with trifluoromethanesulfonic acid (TFMS; Edge et al. 1981; Karp et al. 1982), greater than 91% of the carbohydrate is removed. The deglycosylated products migrate as two bands on one-dimensional SDS-PAGE at 31 kd and 34 kd under nonreducing condition or at 35 kd and 38 kd under reducing condition (Sacco et al. 1986). A similar result has been shown by Henderson et al. (1987). Deglycosylation of the unfractionated zona proteins with TFMS results in the production of five discrete bands on one-dimensional SDS-PAGE with apparent molecular weights of 66, 52, 36, 32, and 16 kd. Two of these (at 36 kd and 32 kd) probably corresponds to the 38-kd and 35-kd products reported by Sacco et al. (1986), respectively. These facts suggests that the 55-kd fraction is composed of two distinct proteins having no interchain disulfide bond, although fractionation of them in original states has not been accomplished so far.

Recently, two research groups have presented a chemical characterization of these proteins. Yurewicz et al. (1987) digested the purified 55-kd fraction with *Escherichia freundii* endo-β-galactosidase to cleave the lactosaminoglycan moieties of the glycoproteins and applied the digests (EBGD) to reverse-phase HPLC using a column of Vydac C_4 equilibrated with 0.1% trifluoroacetic acid. With an increase in the concentration of acetonitrile in the eluate, two peaks containing EBGDβ and EBGDα were eluted successively, and molecular weights of these partially deglycosylated components were estimated to be 46 kd for EBGDα and 43 kd for EBGDβ in nonreduced state. By SDS-PAGE Hedrick and Wardrip (1987) also separated the chemically deglycosylated 55 kd protein with TFMS into 36-kd and 40-kd components in nonreduced state or 37-kd and 40-kd components in reduced state and designated these as 55-kd α and 55-kd β protein, respectively. Attention is necessary not to confuse the designation by two groups: α-glycoprotein, the original state of EBGDα, corresponds, to 55-kd β, while β-glycoprotein, the original state of EBGDβ, corresponds to 55-kd α (Fig. 2).

Compositional Analysis. Amino acid and monosaccharide compositions of the fractionated components expressed as residues per mole of protein are summarized in Table 2. The molar ratio of monosaccharide to amino acids of the 65-kd component (0.101) is higher than that of the 25-kd component (0.068). The sum of the amino acid and carbohydrate of the 65-kd and 25-kd components is almost the same as that of the 90-kd component, supporting the view that 25-kd and 65-kd components are derived from the 90-kd component. The results of amino acid analyses of the 55-kd fraction and its components by Yurewicz et al. (1987) shown in Table 2 are generally consistent with those by Hedrick and Wardrip (1987). No significant difference is found in the amino acid composition of two components, EBGDα and EBGDβ, except for tyrosine. The molar ratio of monosaccharide to amino acid of the 55-kd fraction amounts to 0.25. Galactose and *N*-acetylglucosamine as well as sialic acid are selectively removed by treatment with endo-β-galactosidase. From the results of carbohydrate analyses of EBGDα and EBGDβ, Yurewicz et al. (1987) estimated that α-glycoprotein contains four N-linked and three O-linked oligosaccharides, and that β-glycoprotein contains four N-linked and six O-linked oligosaccharides. Nakano et al. (1987b) have detected xylose and *N*-glycolylneuraminic acid in the 55-kd fraction. Czuppon and Dietl (1985) have also found unusual sialic acids, 5-*N*-glycolylneuraminic acid and 5-N-acetyl-7, 8, 9-tri-O-acetylneuraminic acid, besides 5-N-acetylneuraminic acid in a component

Table 2. Amino acid and monosaccharide compositions of porcine zona proteins

	Residues per mole protein[a]					
	90-kd[b] (PZP2)	65-kd[b]	25-kd[b]	55-kd[c] (PZP3)	46-kd[c] (EBGDα)	42-kd[c] (EBGDβ)
Asx	50.2	39.9	10.0	27.3	26.9	27.5
Thr	30.5	27.1	8.0	24.6	22.7	22.1
Ser	49.0	41.8	9.1	29.6	31.5	25.6
Glx	59.2	47.6	9.8	30.6	29.7	33.3
Pro	24.4	20.6	5.7	28.6	35.0	26.9
Gly	56.3	46.3	9.6	18.3	19.2	17.6
Ala	28.8	23.4	6.4	21.3	23.8	21.5
½ Cys	14.2	9.0	2.7	13.0	15.7	11.2
Val	29.7	28.6	6.5	29.0	33.6	30.4
Met	10.3	6.6	2.7	3.0	3.8	3.2
Ile	18.4	14.0	3.9	10.3	12.2	9.0
Leu	43.0	38.5	8.5	32.0	31.1	29.8
Tyr	14.5	15.7	2.7	8.0	10.5	2.9
Phe	18.1	14.7	4.3	14.6	12.2	13.8
Trp	28.3	11.9	6.7	nd	nd	nd
Lys	27.9	21.6	3.4	11.7	11.9	12.2
His	12.6	12.3	2.3	11.7	10.8	12.8
Arg	22.7	18.0	4.7	19.0	18.2	19.5
Man	9.1	9.3	1.0	9.3	12.1	10.9
Fuc	2.1	2.0	0.5	4.0	4.2	4.5
Gal	13.8	12.1	2.2	28.6	10.8	12.1
GlcNAc	20.0	19.6	3.4	31.3	18.5	18.9
GalNAc	0.0	0.0	0.0	4.3	3.0	6.0
Sialic acid	1.6	1.4	0.2	5.7	1.7	1.9
pI range[d]	4.3-6.4	4.3-6.6	3.8-6.5	3.8-6.7		

PZP, porcine zona protein; EGBD, endo-β-galactosidase digest.

[a] Calculations are based on the molecular weights of the deglycosylated polypeptides: 90 kd, 68000; 65 kd, 56000; 25 kd, 15000; 55 kd, 36500; 46 kd, 38000; and 42 kd, 35000 (Hedrick and Wardrip 1987; Yurewicz et al. 1987).

[b] 65-kd component and 25-kd component constitute N- and C-terminal region of 90-kd protein, respectively.

[c] Calculated from the values reported by Yurewicz et al. (1987). 55-kd proteins are separated into two components (46-kd and 42-kd) after digestion with endo-β-galactosidase.

[d] In iodinated and reduced state (Hedrick and Wardrip 1986a).

of PZP fractionated by HPLC. Separation of the 55-kd fraction into two components in native state without removal of carbohydrates will surely elucidate more detailed features of these predominant fractions of PZP.

Zona Proteins of Other Mammals

Hamster zona proteins in reduced state are resolved into two components of apparent molecular weights of 80 kd and 240 kd by SDS-PAGE. By a second electrophoresis in the same direction without SDS a third band of an apparent molecular weight of 150 kd appears (Ahuja and Bolwell 1983). Three major families of zona protein from human zona pellucida (ZP1, 80–120 kd; ZP2, 73 kd; ZP3, 59–65 kd) and rabbit (ZP1, 100–118 kd; ZP2, 83–110 kd; ZP3, 80–92 kd), and four major families from squirrel monkey (ZP1, 63–78 kd; ZP2, 63–70 kd; ZP3, 47–51 kd; ZP4, 43–47 kd) have been separated by two-dimensional SDS-PAGE under reducing condition. A similar result was obtained in rabbit zona proteins by Dunbar et al. (1981). The zonae of human, squirrel monkey, and rabbit have been suggested to contain a component which possesses antigenic determinants similar to those of PPZA (Sacco et al. 1981).

Characterization of the Fractionated Zona Proteins

Biosynthesis of Zona Proteins

Whether zona proteins are synthesized in the oocyte or in surrounding follicle cells has long been debated. Bleil and Wassarman (1980b) conducted incorporation experiments of [^{35}S]methionine and [^{3}H]fucose into the glycoproteins of follicle-free denuded oocytes and demonstrated that mouse zona glycoproteins (ZP1, ZP2, and ZP3) originate from the oocyte itself rather than from follicle cells. These three glycoproteins, synthesized at 0.4 pg per oocyte per hour, are sulfated during oocyte growth (Shimizu et al. 1983). Protein synthesis and the formation of zona matrix stop before ovulation, and the zona maintains its biological functions that operate after ovulation (Shimizu et al. 1983; Fléchon et al. 1984).

By pulse-chase experiments, experiments using either enzymes or chemicals that remove N- and/or O-linked oligosaccharide from glycoproteins, and experiments using tunicamycin to prevent *N*-linked glycosylation of nascent proteins, Wassarman and his collegues (Greve et al. 1972; Roller and Wassarman 1983). Revealed that MZP2 (120 kd) is synthesized in the growing oocyte as a 81-kd polypeptide chain to which six N-linked oligosaccharides are bound, giving rise to a ZP2 precursor (91 kd). Finally, mature ZP2 is derived from this precursor by the processing of *N*-linked oligosaccharides. In the presence of tunicamycin ZP2 is secreted into zona pellucida as 81-kd and 76-kd glycoproteins at about one-sixth the rate for mature ZP2 in the absence of the antibiotic, suggesting that N-linked oligosaccharides are necessary for normal secretion of ZP2 (Greve et al. 1982; Roller and Wassarman 1983).

On the other hand, mouse ZP3 (83 kd) is synthesized at first as a 44-kd polypeptide chain to which three or four high mannose-type N-linked oligosaccharides are bound, resulting in a 53-kd or 56-kd glycopeptide. These precursors are then processed to the complex-type, and O-linked oligosaccharides are bound to the nascent ZP3 before its secretion. In the presence of tunicamycin ZP3 is secreted into zona pellucida as a heterogeneous species (51 kd) at about three times the rate for mature ZP3 in the absence of tunicamycin (Roller and Wassarman 1983; Wassarman et al. 1985).

Wolgemuth et al. (1984) suggested that the formation of zona pellucida varies among species, since the molecular and biological properties of the zona of various mammals differ from each other. By experiments using antibodies raised against rabbit zona protein fractions they demonstrated that the follicular cells adjacent to the oocyte contribute to the formation of the zona. That is, some of the zona proteins are secreted from the oocyte and others are synthesized by follicular cells in rabbit. A similar view has been described in rat zona pellucida by Takeuchi and Takeuchi (1985). Using an immunohistochemical method, Léveillé et al. (1987) indicated that the native immunogenic material of hamster zona pellucida is synthesized in the oocyte itself but not in the granulosa cells. Thus, detailed investigation on the biosynthesis of zona proteins has been restricted to the mouse zona. Further analyses on other species is necessary to make clear the site of zona synthesis.

Primary Structure of Zona Proteins

Ringuette et al. (1986) isolated cDNA clones coding for mouse ZP3 from ovaries and found that the ZP3 gene is transcribed as a 1.7-kb poly(A) mRNA. Its expression is developmentally regulated in the oocyte (Philpott et al. 1987). As the result of sequencing of the portions of the pZP-3.1 cDNA insert, the codes of an open reading frame between the NciI and FnuD2 sites are found to coincide exactly with the amino acid sequence of an internally cleaved peptide which is obtained by digestion of MZP3 with *Staphylococcus aureus* V-8 protease. This peptide is purified by an anti-ZP3 monoclonal antibody affinity column, and the first 20 amino acid residues from its N terminus have been determined as shown in Fig. 3 A by gas-phase microsequencing. The 12th residue of asparagine from the N terminus is not detected in amino acid sequencing, although its codon, AAC, is found in the corresponding cDNA clone. Since the asparaginyl residue associating N-linked oligosaccharide occurs usually as Asn-X-Ser, this missing asparagine probably binds to N-linked oligosaccharide (Fig. 3 A). It is of interest that the peptide trapped by the affinity column of an anti-ZP3 monoclonal antibody contains one of the three or four N-linked sugar chains in ZP3.

Recently, Yurewicz et al. (1987) determined the N-terminal amino acid sequences of the 55-kd PZP, α- and β-glycoprotein, as shown in Fig. 3 B, C. Although most of the N terminus of β-glycoprotein is blocked, this protein is judged by electrophoresis to be homogeneous. The sequence shown in Fig. 3 C is the result of analysis of its minor fraction having an unblocked N terminus. These two N-terminal sequences are different from each other, and little similarity is found in the peptide maps of the tryptic digests of the two proteins. From these results together with the differences in the immunoreactivity with rabbit polyclonal anti-ZP3 serum, 55-kd fraction is suggested to be composed of two structurally and immunologically distinct proteins produced from different genes. Carbohydrate analysis suggests that α-glycoprotein contains four N-linked and three O-linked oligosaccharides, while β-glycoprotein has four N-linked and six O-linked oligosaccharides. Amino acid sequences indicate that no N-linked sugar chain is present in the N-terminal region of either protein.

Hedrick and Wardrip (1987) derived similar results in analyses of the deglycosylated 37-kd (55-kd α) and 40-kd (55-kd β) proteins from the 55-kd fraction. They determined the N-terminal amino acid sequence of the 55-kd β protein corresponding to the α-glycoprotein by Yurewicz et al. as follows: H_2N-Asp-Val-Pro-Thr-Ile-Gly-Leu-Ser-X-Ala-Pro-Thr-. Only the N-terminal sequence to the seventh

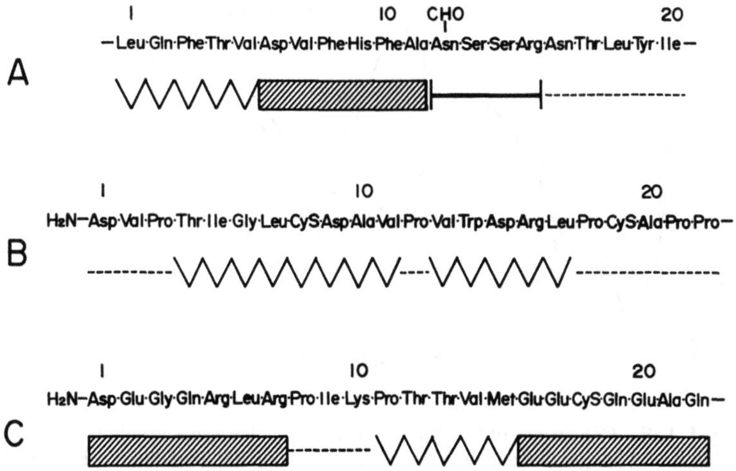

Fig. 3 A–C. Partial primary structure and its predicted secondary structure of mouse and porcine zona proteins. **A** An internal peptide isolated by anti-ZP3 monoclonal antibody affinity chromatography from a V-8 protease digest of mouse ZP3 (Ringuette et al 1986). The 12th residue, Asn, is inferred to be glycosylated. **B** N-terminal region of α-glycoprotein of porcine zona pellucida (Yurewicz et al. 1987). **C** N-terminal region of β-glycoprotein of porcine zona pellucida (Yurewicz et al 1987). The predicted secondary structures by the method of Chou and Fasman (1978) are represented by: *jagged lines,* β-sheet; *shaded bars,* helical structure; *straight line,* β-turn; *broken lines,* coil

residue is identical in the results of these two groups. More detailed peptide analysis is necessary to elucidate the correct amino acid sequence. The finding that another 55-kd component, 55-kd α protein, has a blocked N terminus is consistent with that by Yurewicz et al. (1987).

The N termini of 90-kd and 65-kd components are blocked, while the 25-kd component has an N-terminal sequence of H_2N-Ile-Val-Asn-Gln-Leu-Leu-Asn-X-Ala-Ala-. As described above, these 65-kd and 25-kd components, which are released from the 90-kd component under reducing condition, correspond to N- and C-terminal regions of the 90-kd component, respectively (Fig. 2). Hedrick and Wardrip (1987) have also shown that the 16-kd glycopeptide obtained by digestion of the 65-kd component with V-8 protease gives the sequence of H_2N-X-Gly-Ile-Leu-Phe-Arg-. The same peptide is obtained by direct hydrolysis of the 90-kd component.

The complete amino acid sequences of zona protein, particularly of MZP3 and 55-kd PZP, will soon be determined by cloning techniques or direct protein sequencing.

Secondary Structure of Zona Proteins

A circular dichroic (CD) spectrum of unfractionated PZP in neutral solution shows the β-structure (Nakano et al. 1984). Figure 4 gives CD spectra of three families of PZP obtained by HPLC. In PBS the CD spectra indicate that β-structure is predominant in each family, but that the wavelength of the negative band

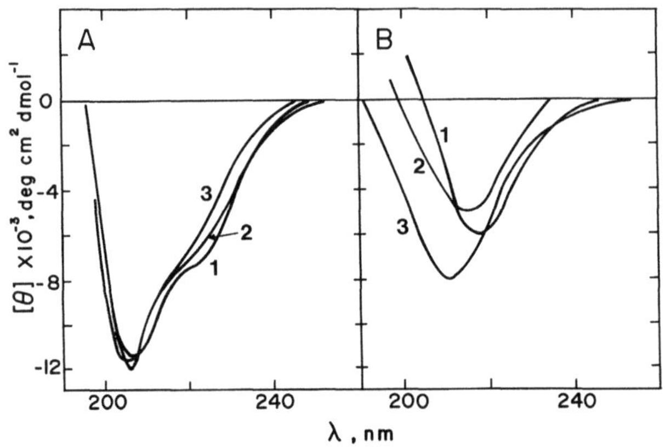

Fig. 4 A, B. CD spectra of the three families of porcine zona protein fractionated by HPLC on a column of TSK gel G3000SW (Nakano et al 1987b). Measurements were performed at 25 °C in 3.5 m*M* SDS, 100 m*M* sodium sulfate, 50 m*M* sodium phosphate, pH 7.0 **(A)** and in PBS, pH 7.4 **(B)**. *1*, PZP1; *2*, PZP2; *3*, PZP3

of PZP3 (55-kd) is shorter and its magnitude larger than those of PZP1 and PZP2. This may be due to a higher content of β-turn in PZP3 than in other two families, since the profile of the typical β-turn has a strong negative band below 190 nm (Chang et al. 1978). Wu and Yang (1978) reported that surfactants provide a proteinaceous environment if the peptide has a structure-forming potential. In the presence of SDS a helical structure is partially induced in the three families but the β-structure is substantially retained (Fig. 4A), indicating that these zona glycoproteins have a structure-forming potential for the β-structure.

Schematic diagrams of the secondary structure predicted by Chou and Fasman's procedure (Chou and Fasman 1978) of a digest of MZP3 with V-8 protease and N-terminal region of porcine glycoproteins are shown in Fig. 3. In the peptide of MZP3 the helical region (residue numbers 6–11) is predicted alternatively for β-structure. In general, the carbohydrate chains associate in β-turn regions which are favorable for inclusion of glycosylated sequences (Hughes 1983). The 12th asparaginyl residue, which is thought to bind to N-linked oligosaccharide, is also present in the β-turn region. On the other hand, the N-terminal region of porcine α-glycoprotein is rich in β-structure. This region is relatively abundant of hydrophobic amino acids, and the hydropathy profile (Kyte and Doolittle 1982) of this region shows a positive pattern, suggesting its being situated in the interior portions of the protein (not shown). In contrast, β-glycoprotein has a relatively hydrophilic N-terminal region. Ascertaining whether the two cysteine residues in α-glycoprotein (numbers 8 and 19) and the one residue in β-glycoprotein (number 18) are present in reduced or oxidized form in the native state will give some information as to the architecture of the zona, since these glycoproteins have only intrachain disulfide bonds.

Supramolecular Structure of Zona Pellucida

Ultrastructural investigation by transmission electron microscopy reveals that the zona pellucida consists of filamentous and particulate materials in pig (Dietl and Czuppon 1984), rat (Takeuchi and Takeuchi 1985), and mouse (Barańska et al. 1975; Greve and Wassarman 1985). The external layer of the mouse zona is found to have a thinner and denser structure than the internal layer (Barańska et al. 1975). In porcine zona a more tightly packed micellar structure has been observed in the internal layer than in the external layer (Dietl and Czuppon 1984). The different properties between the external and internal layers or surfaces are also shown in lectin binding (Nicolson et al. 1975) and in sperm binding (Swenson and Dunbar 1982).

East and Dean (1984) investigated the distribution of MZP using monoclonal antibody to these proteins. ZP2 is present throughout the zona structure, and no evidence of localization is observed. They also found that administration of the monoclonal antibodies, which are not specific for the sperm receptor (see below), inhibits fertilization; this suggests that antibodies to the zona block sperm access by steric hindrance (East et al. 1984). Monoclonal antibodies to porcine zona have also been produced (Isojima et al. 1981, 1984; Drell and Dunbar 1984); porcine zona is ascertained to have cross-reactive antigenicity with human zona (Koyama et al. 1985; Mori et al. 1985). These monoclonal antibodies will give valuable information on the topology of the zona proteins.

In 1985 Greve and Wassarman represented a model of mouse zona pellucida composed of three kinds of glycoproteins. Their "beads-on-a-string" model, with beads of 9.5 nm in diameter located every 17 ± 4.6 nm along the axis of a filament, is based on electron micrographs of negatively stained preparations. The filaments appear to be interconnected by ZP1, forming a three-dimensional zona matrix. Treatment of the zona with chymotrypsin or dithiothreitol causes the cleavage of ZP1, giving rise to solubilization of the zona and disruption of interconnections between individual filaments, but heterologous interactions between ZP2 and ZP3 are maintained. These observations suggest that the zona pellucida has a highly organized and repeated supramolecular structure, although its molecular basis remains to be determined. The CD profile of the extensively trypsin-solubilized porcine zona also suggests the presence of structural repeats in zona architecture (Nakano et al. 1987a). Trypsin seems preferentially to cleave β-turn regions of zona proteins, resulting in the appearance in SDS-PAGE of many discrete bands. Differences in solubilization rate between PZP by acrosin (Urch et al. 1985), trypsin, or pronase (Nakano et al. 1987a) may be attributable to this particular structure of the zona.

The porcine zona contains 33 ng protein (Dunbar et al. 1980) with an average molecular weight about 50 kd. All of these proteins adopting a fully extended peptide form correspond to a pattern of about 4×10^{11} molecules with 1.6×10^{2} nm length, because the peptide unit of an extended chain is 0.36 nm long (Corey and Pauling 1953). Accordingly, the total volume occupied by these proteins is no more than 1×10^{13} nm^3. Since the porcine ovum has a diameter of about 157 μm (Dunbar et al. 1980), and the thickness of its zona is 16 μm (Dunbar 1983), the zona has a capacity of 1×10^{15} nm^3. Protein and carbohydrate account for more than 90% of the material of porcine zona, and the weight ratio of protein to carbohydrate is about 3.0 (Dunbar et al. 1980). The volume occupied by carbohydrate moiety is estimated to be at most 5×10^{12} nm^3. Accordingly, there must be much space between peptide chains when these constituents are packed into the zona. A similar estimation is valid when the zona proteins form a β-structure, the peptide unit of which is 0.33 nm long (Pauling and Corey 1951). At all events, the zona has a highly porous structure in which carbohydrates are sparcely distributed (Nicolson et al. 1975). Water molecules probably hydrate to these zona glycoproteins, forming a fenestrate and latticelike structure of the zona (Phillips and Shalgi 1980), and this native supramolecular structure is changed in hypotonic solution (Nakano et al. 1987a).

Biological Function of Fractionated Zona Proteins

In the initial stages of fertilization sperm must bind to and penetrate the zona pellucida before fusion with the plasma membrane of the ovulated egg. In mouse as many as 1500 sperm can be bound to a single egg in vitro a very high concentration of sperm. On binding, sperm and egg are thought to recognize mutually the species-specific receptor in their surfaces. Some antibodies raised against ovarian antigen have been found to be associated with the zona pellucida, forming a precipitate on the egg surface (Dunbar and Shivers 1976). Furthermore, it has been demonstrated that antibodies to the zona block fertilization both in vitro and in vivo, as illustrated in Fig. 5B (Tsunoda and Sugie 1979; Dunbar 1983). More

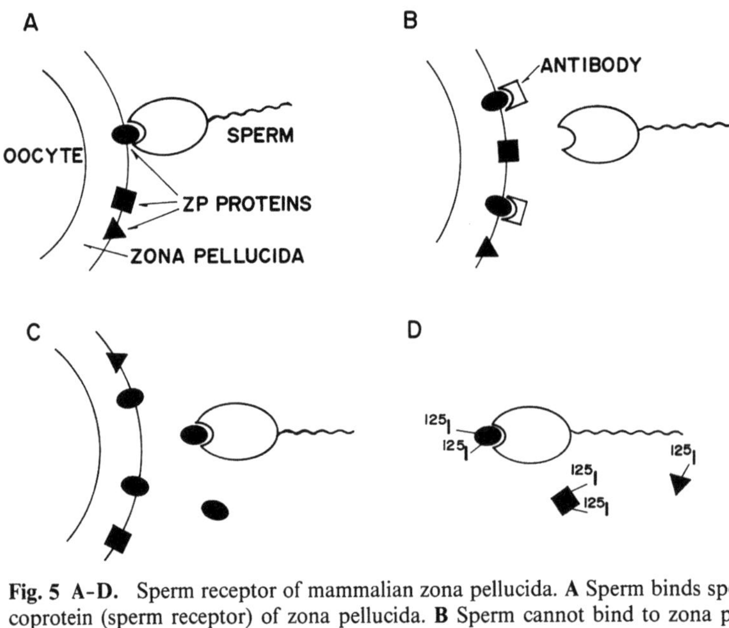

Fig. 5 A–D. Sperm receptor of mammalian zona pellucida. **A** Sperm binds specifically to a glycoprotein (sperm receptor) of zona pellucida. **B** Sperm cannot bind to zona pellucida of ovum treated with the antibody against sperm receptor. **C** Sperm treated with sperm receptor cannot bind to zona pellucida (competition assay). **D** ^{125}I-Labeled sperm receptor is specifically bound to sperm (radiobinding assay)

recently East et al. (1984, 1985) reported that monoclonal antibodies specific for mouse ZP3 inhibit in vivo and in vitro fertilization, and that monoclonal antibodies specific for ZP2 are effective in blocking fertilization in vivo.

By "competition assay", as shown in Fig. 5C, Bleil and Wassarman (1980c) demonstrated that one of the three mouse zona glycoproteins, ZP3, has the sperm receptor activity which is responsible for the binding of sperm to the zona of unfertilized eggs but not to that of a fertilized two-cell embryo. ZP3 purified from the zonae of unfertilized eggs also induces acrosome reaction in vitro; ZP1 and ZP2 from unfertilized eggs as well as ZP3 from embryos have little ability for the induction of acrosome reaction (Bleil and Wassarman 1983). The concentration of ZP3 in the zona (about 40000 molecules/μm^3) is sufficient to induce the acrosome reaction after sperm binding. The sperm receptor activity of ZP3 has been found to be associated with the O-linked oligosaccharides that have an apparent molecular weight of 3.9 kd and are released with mild alkaline hydrolysis and reduction of ZP3. On the other hand, the polypeptide chain of ZP3 plays a role in inducing the acrosome reaction (Florman et al. 1984; Florman and Wassarman 1985; Wassarman et al. 1985; Wassarman 1987).

The sperm receptor of the zona of other mammals than mouse has been recognized only in porcine zona. Dietl et al. (1983c) developed the "radiobinding assay" shown in Fig. 5D and found that one of the ^{125}I-labeled PZP, which are fractionated by HPLC, binds specifically to the porcine sperm. As a result of binding only the sperm treated with ^{125}I-labeled 38-kd protein is found to have the radioactivity. However, no fraction isolated by other research groups (Table 2) corresponds to this 38-kd protein, which has alanine as N terminus and contains no cysteine or methionine. Although acidic glycoprotein (PPZA) with an apparent molecular weight of 58 kd is suggested to have receptor activity by competition assay (Sacco et al. 1984), this fraction is not homogeneous (Yurewicz et al. 1983). Berger et al. (1986) reported that both the 90-kd component the 55 kd component inhibits sperm binding to the zona in competition assay. Thus, it is hard to say that porcine sperm receptor has been isolated so far.

Bleil and Wassarman (1986) have adopted a method similar to radiobinding assay in mouse and found that ^{125}I-labeled ZP3 is bound to the acrosome cap region of acrosome-intact mouse sperm but not to the heads of acrosome-reacted sperm. In contrast, ^{125}I-labeled ZP2 is preferentially bound to acrosome-reacted sperm. The ZP2 component has been considered to play a role in polyspermy blocking since a specific proteolysis of ZP2 is observed following fertilization (Bleil et al. 1981). Establishment of fertilization in mouse induces an increase in resistance to proteases and solubilizing agents (Schmell and Gulyas 1980), as in the case of sea urchin (Foerder and Shapiro 1977). Although this phenomenon of hardening is thought to be attributable to the cross-links of tyrosine residues and the formation of a fertilization membrane by a cortical granule peroxidase (ovoperoxidase), the formation of cross-linking of MZP has not been confirmed (Wassarman 1987).

Recently Nakano et al. (1987a) noted the production of high molecular weight products by hydrolysis of porcine zona by proteases in isotonic solution. However, the property of these products, which are not dissociated even in SDS solution, has not been clarified. In pig the zonae from ovulated eggs and embryos possess the additional glycoproteins which originate from the oviduct and bind firmly to the zona (Brown and Cheng 1986). These proteins accept hydrolytic processing probably by the enzymes derived from cortical granules at fertilization (Hedrick et al. 1987).

Concluding Remarks

Mammalian zona pellucida is a scientifically attractive substance which exhibits several biological functions and relatively few components. In order to understand its functions at the molecular level the development of better procedures for purification of zona materials and for analysis of its individual components is essential. Also necessary are more advanced methods for detecting slight molecular changes in zona components and zona architecture in the performance of these biological functions. Elucidation of the true function of zona pellucida and clarification of the mechanism of mutual recognition between zona and sperm will make a fresh development in cell biology.

Acknowledgment. I am grateful to Professor Tohru Tobita and Ms. Yumiko Hatanaka for their valuable comments in preparing this manuscript.

References

Ahuja KK, Bolwell GP (1983) Probable asymmetry in the organization of components of the hamster zona pellucida. J Reprod Fertil 69: 49-55

Aitken RJ, Richardson DW (1981) Measurement of the sperm binding capacity of the mouse zona pellucida and its use in the estimation of anti-zona antibody titres. J Reprod Fertil 63: 295-307

Aitken RJ, Holme E, Richardson DW, Hulme M (1982) Properties of intact and univalent (Fab) antibodies raised against isolated, solubilized, mouse zonae pellucidae. J Reprod Fertil 66: 327-334

Barańska W, Konwiński M, Kujawa M (1975) Fine structure of the zona pellucida of unfertilized egg cells and embryos. J Exp Zool 192: 193-202

Berger T, Goldhawk DE, Wardrip NJ, Davis A, Hedrick JL (1986) Interaction of zona pellucida components with porcine sperm. J Cell Biol 103: 368a

Bleil JD, Wassarman PM (1980a) Structure and function of the zona pellucida: identification and characterization of the proteins of the mouse oocyte's zona pellucida. Dev Biol 76: 185-202

Bleil JD, Wassarman PM (1980b) Synthesis of zona pellucida proteins by denuded and follicle-enclosed mouse oocytes during culture in vitro. Proc Natl Acad Sci USA 77: 1029-1033

Bleil JD, Wassarman PM (1980c) Mammalian sperm-egg interaction: Identification of a glycoprotein in mouse egg zonae pellucidae possessing receptor activity for sperm. Cell 20: 873-882

Bleil JD, Wassarman PM (1983) Sperm-egg interactions in the mouse: sequence of events and induction of the acrosome reaction by a zona pellucida glycoprotein. Dev Biol 95: 317-324

Bleil JD, Wassarman PM (1986) Autoradiographic visualization of the mouse egg's sperm receptor bound to sperm. J Cell Biol 102: 1363-1371

Bleil JD, Beall CF, Wassarman PM (1981) Mammalian sperm-egg interaction: fertilization of mouse eggs triggers modification of the major zona pellucida glycoprotein, ZP2. Dev Biol 86: 189-197

Brauer MM, Sotelo JR (1983) Large scale isolation of zonae pellucidae from ovarian oocytes of mice. Experientia 39: 420-421

Brown CR, Cheng WTK (1985) Limited proteolysis of the porcine zona pellucida by homologous sperm acrosin. J Reprod Fertil 74: 257-260

Brown CR, Cheng WKT (1986) Changes in composition of the porcine zona pellucida during development of the oocyte to the 2- to 4-cell embryo. J Embryol Exp Morphol 92: 183-191

Chang CT, Wu CSC, Yang JT (1979) Circular dichroic analysis of protein conformation: inclusion of the β-turns. Anal Biochem 91: 13-31

Chou PY, Fasman GD (1978) Empirical predictions of protein conformation. Annu Rev Biochem 47: 251-276

Corey RB, Pauling L (1953) Fundamental dimensions of polypeptide chains. Proc R Soc Lond [Biol] 141: 10-20

Czuppon AB, Dietl J (1985) Biochemical characterization of five glycoprotein molecules of the porcine zona pellucida. Biochem Int 11: 21-34

Dietl J, Czuppon AB (1984) Ultrastructural studies of the porcine zona pellucida during the solubilization process by Li-3,5-diiodosalicylate. Gamete Res 9: 45-54

Dietl J, Czuppon AB, Königsmann W (1983a) Separation of porcine zona pellucida components by high-performance liquid chromatography on styro-gel protein columns. J Chromatogr 275: 423-427

Dietl, J, Czuppon A, Mettler L (1983b) Characterization of porcine zona pellucida antigens by immunoaffinity chromatography and by high-pressure liquid chromatography. Am J Reprod Immunol 4: 116-121

Dietl J, Czuppon A, Weichert K, Mettler L (1983c) Identification of a sperm receptor-glycoprotein from the porcine zona pellucida. Hoppe Seylers Z Physiol Chem 364: 261-267

Drell DW, Dunbar BS (1984) Monoclonal antibodies to rabbit and pig zonae pellucidae distinguish species-specific and shared antigenic determinants. Biol Reprod 30: 445-457

Drell DW, Wood DM, Bundman D, Dunbar BS (1984) Immunological comparison of antibodies to porcine zonae pellucidae in rats and rabbits. Biol Reprod 30: 435-444

Dunbar BS (1983) Morphological, biochemical, and immunochemical characterization of the mammalian zona pellucida. In: Hartmann JF (ed) Mechanism and control of animal fertilization. Academic, New York, pp 139-175

Dunbar BS, Raynor BD (1980) Characterization of porcine zona pellucida antigens. Biol Reprod 22: 941-954

Dunbar BS, Shivers CA (1976) Immunological aspects of sperm receptors on the zona pellucida of mammalian eggs. Immunol Commun 5: 375-385

Dunbar BS, Wardrip NJ, Hedrick JL (1980) Isolation, physicochemical properties, and macromolecular composition of zona pellucida from porcine oocytes. Biochemistry 19: 356-365

Dunbar BS, Liu C, Sammons DW (1981) Identification of the three major proteins of porcine and rabbit zonae pellucidae by high resolution two-dimensional gel electrophoresis: comparison with serum, follicular fluid, and ovarian cell proteins. Biol Reprod 24: 1111-1124

Dunbar BS, Dudkiewicz AB, Bundman DS (1985) Proteolysis of specific porcine zona pellucida glycoproteins by boar acrosin. Biol Reprod 32: 619-630

East IJ, Dean J (1984) Monoclonal antibodies as probes of the distribution of ZP-2, the major sulfated glycoprotein of the murine zona pellucida. J Cell Biol 98: 795-800

East IJ, Mattison DR, Dean J (1984) Monoclonal antibodies to the major protein of the murine zona pellucida: effects on fertilization and early development. Dev Biol 104: 49-56

East IJ, Gulyas BJ, Dean J (1985) Monoclonal antibodies to the murine zona pellucida protein with sperm receptor activity: Effects on fertilization and early development. Dev Biol 109: 268-273

Edge ASB, Faltynek CR, Hof L, Reichert LE Jr, Weber P (1981) Deglycosylation of glycoproteins by trifluoromethanesulfonic acid. Anal Biochem 118: 131-137

Fairbanks G, Steck TL, Wallach DFH (1971) Electrophoretic analysis of the major polypeptides of the human erythrocyte membrane. Biochemistry 10: 2606-2617

Fléchon JE, Pavlok A, Kopečný V (1984) Dynamics of zona pellucida formation by the mouse oocyte. An autoradiographic study. Biol Cell 51: 403-406

Florman HM, Wassarman PM (1985) O-Linked oligosaccharides of mouse egg ZP3 account for its sperm receptor activity. Cell 41: 313-324

Florman HM, Bechtol KB, Wassarman PM (1984) Enzymatic dissection of the functions of the mouse egg's receptor for sperm. Dev Biol 106: 243-255

Foerder CA, Shapiro BM (1977) Release of ovoperoxidase from sea urchin eggs hardens the fertilization membrane with tyrosine crosslinks. Proc Natl Acad Sci USA 74: 4214-4218

Greve JM, Wassarman PM (1985) Mouse egg extracellular coat is a matrix of interconnected filaments possessing a structural repeat. J Mol Biol 181: 253-264
Greve JM, Salzmann GS, Roller RJ, Wassarman PM (1982) Biosynthesis of the major zona pellucida glycoprotein secreted by oocytes during mammalian oogenesis. Cell 31: 749-759
Gwatkin RBL, Andersen OF, Williams DT (1980) Large scale isolation of bovine and pig zonae pellucidae: chemical, immunological, and receptor properties. Gamete Res 3: 217-231
Hartmann JF, Gwatkin RBL (1971) Alteration of sites on the mammalian sperm surface following capacitation. Nature 234: 479-481
Hartmann JF, Hutchison CF, Vandlen RL (1982) Modulation of fertilization by the vitellus after interaction with peptides released by hamster sperm-zona pellucida contact. Dev Biol 93: 145-151
Hedrick JL, Wardrip NJ (1981) Microheterogeneity in the glycoproteins of the zona pellucida is due to the carbohydrate moiety. J Cell Biol 91: 177a
Hedrick JL, Wardrip NJ (1986a) Isolation of the zona pellucida and purification of its glycoprotein families from pig oocytes. Anal Biochem 157: 63-70
Hedrick JL, Wardrip NJ (1986b) Proteolysis by collagenase preparations alters the macromolecular composition of the porcine zona pellucida. Biol Reprod 35: 677-682
Hedrick JL, Wardrip NJ (1987) On the macromolecular composition of the zona pellucida from porcine oocytes. Dev Biol 121: 478-488
Hedrick JL, Wardrip NJ, Berger T (1987) Differences in the macromolecular composition of the zona pellucida isolated from pig oocytes, eggs, and zygotes. J Exp Zool 241: 257-262
Henderson CJ, Hulme MJ, Aitken RJ (1987) Analysis of the biological properties of antibodies raised against intact and deglycosylated porcine zonae pellucidae. Gamete Res 16: 323-341
Hughes RC (1983) Glycoproteins. Chapman and Hall, London
Isojima S, Koyama K, Hasegawa A (1981) Production of monoclonal antibody to zona pellucida from porcine oocytes. Acta Obstet Gynaecol Jpn 33: 1995-1996
Isojima S, Koyama K, Hasegawa A, Tsunoda Y, Hanada A (1984) Monoclonal antibodies to porcine zona pellucida antigens and their inhibitory effects on fertilization. J Reprod Immunol 6: 77-87
Karp DR, Atkinson JP, Shreffler DC (1982) Genetic variation in glycosylation of the fourth component of murine complement. J Biol Chem 257: 7330-7335
Koyama K, Hasegawa A, Tsuji Y, Isojima S (1985) Production and characterization of monoclonal antibodies to cross-reactive antigens of human and porcine zonae pellucidae. J Reprod Immunol 7: 187-198
Kyte J, Doolitle RF (1982) A simple method for displaying the hydropathic character of a protein. J Mol Biol 157: 105-132
Laemmli UK (1970) Cleavage of structural proteins during the assembly of the head of bacteriophage T4. Nature 227: 680-685
Léveillé MC, Roberts KD, Chevalier S, Chapdelaine A, Bleau G (1987) Formation of the hamster zona pellucida in relation to ovarian differentiation and follicular growth. J Reprod Fertil 79: 173-183
Loewenstein JE, Cohen AI (1964) Dry mass, lipid content, and protein content of the intact and zona-free mouse ovum. J Embryol Exp Morphol 12: 113-121
Menino AR Jr, Wright RW Jr (1979) Characterization of porcine oocyte zonae pellucidae by polyacrylamide gel electrophoresis. Proc Soc Exp Biol Med 160: 449-452
Mori T, Kamada M, Yamano S, Kinoshita T, Kano K, Mori T (1985) Production of monoclonal antibodies to porcine zona pellucida and their inhibition of sperm penetration through human zona pellucida in vitro. J Reprod Immunol 8: 1-11
Nakano M, Otsuka M, Akama K, Tobita T (1984) Predominance of β-structure in solubilized zona pellucida from porcine ova. Biochem Int 9: 39-43
Nakano M, Hatanaka Y, Tobita T (1987a) Solubilization of porcine zonae pellucidae by trypsin and pronase. Biochem Int 14: 425-433
Nakano M, Hatanaka Y, Sawai T, Kobayashi N, Tobita T (1987b) Fractionation of glycoproteins from porcine zonae pellucidae into three families by high-performance liquid chromatography. Biochem Int 14: 417-423
Nicolson GL, Yanagimachi R, Yanagimachi H (1975) Ultrastructural localization of lectin-bind-

ing sites on the zonae pellucidae and plasma membranes of mammalian eggs. J Cell Biol 66: 263-274
Noda Y, Mori T, Takai I, Kohda H, Nishimura T (1981) A new method for preparation of pure zonae pellucidae in large quantities from porcine ovaries. J Reprod Immunol 3: 147-156
Noda Y, Kohda H, Takai I, Hayashi S, Shimada H, Mori T, Tojo S (1983) Characterization of glycoproteins isolated from porcine zonae pellucidae. J Reprod Immunol 5: 161-172
O'Farrell PH (1975) High resolution two-dimensional electrophoresis of proteins. J Biol Chem 250: 4007-4021
Oikawa T (1978) A simple method for the isolation of a large number of ova from pig ovaries. Gamete Res 1: 265-267
Pauling L, Corey RB (1951) Configurations of polypeptide chains with favored orientations around single bonds: two new pleated sheets. Proc Natl Acad Sci USA 37: 729-740
Phillips DM, Shalgi RM (1980) Surface properties of the zona pellucida. J Exp Zool 213: 1-8
Philpott CC, Ringuette MJ, Dean J (1987) Oocytespecific expression and developmental regulation of ZP3, the sperm receptor of the mouse zona pellucida. Dev Biol 121: 568-575
Ringuette MJ, Sobieski DA, Chamow SM, Dean J (1986) Oocyte-specific gene expression: molecular characterization of a cDNA coding for ZP-3, the sperm receptor of the mouse zona pellucida. Proc Natl Acad Sci USA 83: 4341-4345
Roller RJ, Wassarman PM (1983) Role of asparaginelinked oligosaccharides in secretion of glycoproteins of the mouse egg's extracellular coat. J Biol Chem 258: 13243-13249
Sacco AG (1977) Antigenic cross-reactivity between human and pig zona pellucida. Biol Reprod 16: 164-173
Sacco AG, Yurewicz EC, Subramanian MG, DeMayo FJ (1981) Zona pellucida composition: species cross reactivity and contraceptive potential of antiserum to a purified pig zona antigen (PPZA). Biol Reprod 25: 997-1008
Sacco AG, Subramanian MG, Yurewicz EC (1984) Association of sperm receptor activity with a purified pig zona antigen (PPZA). J Reprod Immunol 6: 89-103
Sacco AG, Yurewicz EC, Subramanian MG (1986) Carbohydrate influences the immunogenic and antigenic characteristics of the ZP3 macromolecule (Mr55000) of the pig zona pellucida. J Reprod Fertil 76: 575-586
Schmell ED, Gulyas BJ (1980) Ovoperoxidase activity in ionophore treated mouse eggs. II. Evidence of the enzyme's role in hardening the zona pellucida. Gamete Res 3: 279-290
Schultz RM, Letourneau GE, Wassarman PM (1979a) Program of early development in the mammal: changes in patterns and absolute rates of tubulin and total protein synthesis during oogenesis and early embryogenesis in the mouse. Dev Biol 68: 341-359
Schultz RM, Letourneau GE, Wassarman PM (1979b) Program of early development in the mammal: changes in the patterns and absolute rates of tubulin and total protein synthesis during oocyte growth in the mouse. Dev Biol 73: 120-133
Shimizu S, Ito M (1986) Glycoproteins of murine zona pellucida from unfertilized eggs and two-cell embryos: comparison of the reactivity to lectins. Biochem Int 13: 335-341
Shimizu S, Ito M, Dean J (1982) Glycoproteins of mouse zona pellucida: analysis of their reactivity to lectins. Biochem Biophys Res Commun 109: 449-454
Shimizu S, Tsuji M, Dean J (1983) In vitro biosynthesis of three sulfated glycoproteins of murine zonae pellucidae by oocytes grown in follicle culture. J Biol Chem 258: 5858-5863
Sullivan R, Bleau G (1985) Interaction of isolated components from mammalian sperm and egg. Gamete Res 12: 101-116
Swenson CE, Dunbar BS (1982) Specificity of sperm-zona interaction. J Exp Zool 219: 97-104
Takeuchi IK, Takeuchi YK (1985) Use of tannic acid for ultrastructural visualization of zonae pellucidae in growing oocytes of rat ovarian follicle. Zool Sci 2: 135-139
Tsunoda Y, Sugie T (1979) Production of hetero-antibody to isolated pig zonae pellucidae. Jpn J Zootech Sci 50: 493-498
Urch UA, Wardrip NJ, Hedrick JL (1985) Limited and specific proteolysis of the zona pellucida by acrosin. J Exp Zool 233: 479-483
Wassarman PM (1983) Oogenesis: synthetic events in the developing mammalian egg. In: Hartmann JF (ed) Mechanism and control of animal fertilization. Academic, New York, pp 1-54
Wassarman PM (1987) The biology and chemistry of fertilization Science 235: 553-560

Wassarman PM, Bleil JD, Cascio SM, LaMarca MJ, Letourneau GE, Mrozak SC, Schultz RM (1981) Programming of gene expression during mammalian oogenesis. In: Jagiello G, Vogel HJ (eds) Bioregulators of reproduction. Academic, New York, pp 119-150
Wassarman PM, Bleil JD, Florman HM, Greve JM, Roller RJ, Salzmann GS, Samuels FG (1985) The mouse egg's receptor for sperm: what is it and how does it work? Cold Spring Harbor Symp Quant Biol 50: 11-19
Wolf DP (1982) The ovum before and after fertilization. In: Zaneveld LJD, Chatterton RT (eds) Biochemistry of mammalian reproduction. Wiley and Sons, New York, pp 231-259
Wolgemuth DJ, Celenza J, Bundman DS, Dunbar BS (1984) Formation of the rabbit zona pellucida and its relationship to ovarian follicular development. Dev Biol 106: 1-14
Wu CSC, Yang JT (1978) Conformation of naturally-occurring peptides in surfactant solution: its relation to the structure-forming potential of amino acid sequence. Biochem Biophys Res Commun 82: 85-91
Yanagimachi R (1977) Specificity of sperm-egg interaction. In: Edidin M, Johnson MH (eds) Immunobiology of gametes. Cambridge University Press, London, pp 255-295
Yurewicz EC, Sacco AG, Subramanian MG (1983) Isolation and preliminary characterization of a purified pig zona antigen (PPZA) from porcine oocytes. Biol Reprod 29: 511-523
Yurewicz EC, Sacco AG, Subramanian MG (1987) Structural characterization of the Mr = 55000 antigen (ZP3) of porcine oocyte zona pellucida. J Biol Chem 262: 564-571

Transfer of Oviductal Proteins to the Zona Pellucida*

G. Bleau and S. St-Jacques

Introduction

Upon entering the oviduct the oocyte begins an episode of isolation from any direct cell-cell interaction with the mother. The pseudopodlike processes of the corona radiata retract, leaving the oocyte floating free within the zona pellucida. The fertilizing sperm is the only cell that is allowed intimate contact with the plasma membrane of the female gamete. Up until implantation all signals to and from the developing embryo must be channelled through the zona pellucida.

The zona (see Dunbar 1983, for review) is a layer of glycoproteins produced by the oocyte itself during follicular development. Under the scanning electron microscope, it presents a fenestrated outer surface while the inner surface has a smoother appearance. Several biological functions have been attributed to the zona, such as insuring species specificity of sperm-egg interaction, participating in the block to polyspermy, and protecting the oocyte and the developing embryo during oviductal transit. Should the zona be prematurely shed or lost, the oocyte as well as the embryo are doomed.

At ovulation the oviducts receive the freshly released eggs. Several crucial events occur within the oviduct, including transport of the male and female gametes, support of their final maturation, fertilization, embryonic development, and transport of the embryo to the uterus. The oviductal fluid provides the optimal milieu for the successful completion of all of these early reproductive events (Mastroianni and Go 1979).

Muscular and ciliary movements are involved in the transport of the gametes and in the even distribution of the oviductal fluid.

The tubal fluid is derived from a combination of transudation from the blood and active secretion by the epithelial secretory cells. It provides the supportive and nutritive medium for fertilization and all subsequent embryonic stages. Around the time of ovulation secretory activity and production of oviductal fluid increase under the influence of steroid hormones. This fluid contains most of the serum proteins (at lower concentration), electrolytes, amino acids, and small metabolites (Hamner and Fox 1969). In addition, many specific glycoproteins that are produced and secreted by the oviduct have been described in several species. Some of these specific glycoproteins become associated with the egg and particularly with

* The research work of the authors is supported by the Medical Research Council of Canada, fonds de la Recherche en Santé du Québec, and fonds pour la formation de Chercheurs et l'Aide à la Recherche.

the zona pellucida. The exact nature of these proteins is under study in several laboratories. Their role, if any, probably pertains to assisting and/or complementing the biological functions of the zona pellucida.

Oviductal Proteins

The data on secretion and hormonal control of specific macromolecules secreted by the oviduct have been reviewed by Oliphant in 1986, and a complete description of the glycoproteins presently known to be secreted by rabbit oviduct is presented.

The rabbit is the animal model which has received by far the most attention. This is mainly due to technical reasons, such as easy access to an open fimbria, oviductal fluid that can be obtained in sufficient amount for analysis, precise and easy timing of ovulation, and easy housing. However, because the rabbit is so unique with respect to the dynamics of egg covering within the oviduct, investigators in this field of research are choosing other animal models which could allow the formulation of more general concepts applicable to animal as well as human reproduction.

The importance and interest generated by the work of Oliphant and collaborators has certainly been an impetus for the study of oviductal proteins in other species. Both from earlier and from recent studies we know that the oviduct or oviductal fluid contains specific components (glycoproteins) in the sheep (Roberts et al. 1976; Sutton et al. 1984), cow (Roberts et al. 1975), gilt (Buhi 1986), hamster (Noske and Daniel 1974), monkey (Mastroianni et al. 1970; Fazleabas and Verhage 1986), and human (Moghissi 1970; Lippes et al. 1983; Matsumoto et al. 1986). These studies suggest that the mammalian oviduct produces and secretes at least one major family of estrus-associated secretory macromolecules absent from the serum. These specific glycoproteins are present in oviductal fluid up until the time when the embryo reaches the uterus, suggesting that they could be involved in fertilization and/or embryonic development. Some of these glycoproteins might represent those that can bind to the oviductal egg. In the reports that are described above, this possibility was not explored. The following section deals specifically with this particular aspect.

Oviductal Proteins on the Egg

Rabbit

Many authors have described the addition of proteins to the ovulated ova and early embryos during their transit in the oviduct. This is most evident and has been known for a long time in the rabbit, where a thick mucin layer is deposited around the egg (review by Greenwald 1969). The thickness of the mucin layer on the ova and cleavage-stage embryos is influenced by the hormonal status of the animal. The dynamic structure of the covering of the rabbit blastocyst has been meticulously described by Denker and Gerdes in 1979. During the blastocyst stages the zona pellucida gradually disappears (dissolution), with the concomitant deposition of a new layer (possibly originating from the trophoblast) called the neozona. The mucin coat becomes thinner as the blastocyst expands and is later covered by an external layer of material (gloiolemma) secreted by the uterus.

The mucoproteins do not adhere in a specific manner to the surface of the zona pellucida; thus, a mucin coat accumulates around foreign objects (glass beads) introduced into the rabbit oviduct (Greenwald 1969). Many investigators have observed the presence of a thin layer of mucin around bovine, porcine, and goat eggs transferred to the rabbit oviduct (Lawson et al. 1972; Hirst et al. 1981; Sirard et al. 1985; Rao et al. 1984; Boland 1984). The amount of mucin deposited on cow embryos was reported to be about one-fifth that on the eggs of recipient does (Sirard et al. 1985) but similar to the amount found on glass beads by Bourdage and Halbert (1984). The mucin coat does not seem to impair the subsequent development of the embryos.

The material of the mucin coat consists of acidic glycoproteins (Braden 1952) of very high molecular weight (Shapiro et al. 1974). The latter authors have analyzed the proteins of rabbit oviductal fluid: under dissociating and reducing conditions a molecular weight of 73-kd was found for the predominant major band (monomer). An antiserum raised against this glycoprotein reacted with the egg coating.

More recently, sulfated glycoproteins have been isolated and purified from rabbit oviductal fluid (Barr and Oliphant 1984; Oliphant and Ross 1982). These proteins are synthesized by the oviductal epithelium, and the monomeric forms have a molecular weight of 71-kd. Now that polyclonal and monoclonal antibodies are available which react with these glycoproteins (Erickson et al. 1986), it will be possible to use immunohistochemical methods with a high resolution to determine whether they are components of the mucin coat. These proteins might be the same as those isolated earlier by Shapiro et al. (1984): they represent the major secretory glycoproteins and have similar molecular weights.

On the basis of the few histochemical and immunological studies on the deposition of glycoproteins forming the mucin coat, it would appear that the glycoproteins do not become integrated into the zona pellucida (Sacco and Shivers 1973a). The mucin layer seems to be simply adhering to the outermost surface of the zona and to develop as a thick layer. The rabbit and marsupials are mammals in which this type of egg covering is known to take place within the oviduct. Other mammalian species have evolved differently with respect to egg processing within the oviduct. It is not excluded, however, that minor secretory glycoproteins of high monomeric molecular weight could also be produced by the rabbit oviduct and subsequently transferred to the zona pellucida.

Mouse

Baranska et al. (1975) described a fine-grained peripheral layer outside the external layer of the zona pellucida of mouse oviductal ova (both unfertilized ova and embryos). They proposed that it could be contributed by the secretions of the oviduct and of the uterus.

More recently Kapur and Johnson (1985) reported on a 215-kd glycoprotei associated with ovulated oocytes and early embryos obtained from the oviduct Preovulatory oocytes did not contain this glycoprotein, but they could acquire i upon in vitro incubation in bursal fluid. The glycoprotein (detected with the use o wheat germ agglutinin) was proposed to originate from the oviductal epithelia cells and to associate secondarily with the ovulated ovum. Evidence was presente that the 215-kd glycoprotein is not a component of the zona pellucida. Hence manually isolated zonae did not yield lectin binding in the high molecular weigh region of the gel characteristic of the glycoprotein. Immunohistochemical tech niques at the electron-microscopic level would help to define the precise site o binding in this species.

Glass (1963) has described the transfer of blood-borne macromolecules to th cytoplasm of oviductal eggs in the mouse. It is thus possible that in this species th oviduct specific glycoproteins could become associated with the ooplasm itsel and not with the zona.

Pig

In addition to the intrinsic zona proteins, Brown and Cheng (1985) have detecte the presence of a 250-kd protein in porcine zona pellucida of eggs recovered fror the oviduct. They proposed that this protein is apparently absorbed from the ovi duct. They observed that the protein can be rapidly hydrolyzed by treating egg with acrosin to generate an acrosin-resistant polypeptide of 114-kd whic remained firmly bound to the zona. It will be most interesting to determin whether one of the oviductal porcine glycoproteins isolated by Buhi (1986) corre sponds to this 250-kd protein.

Hamster

Fox and Shivers (1975) were the first to describe the addition of oviductal compo nents to the hamster zona pellucida. In their indirect immunofluorescence studie they incubated eggs in the presence of anti-uterus sera (absorbed with small inte tine, lung, and liver), followed by the second fluorescein isothiocyanate (FITC labeled antibody. A fluorescent reaction was detected on oviductal eggs an embryos but not on ovarian eggs; the reaction appeared to be concentrated mainl in a band at the periphery of the zona.

Oikawa and Sendai (1984) presented preliminary evidence confirming that glycoprotein is added to the zona pellucida of ovulated eggs during transit in th oviduct. They observed that fluorescein-conjugated *Bandeirae simplicifolia* lecti (FITC-bovine serum albumin, or BSA 1 binds to specific terminal saccharide res dues) binds to oviductal but not to ovarian eggs. When exposed to an ovidu extract, ovarian zonae bind FITC-BSA-1 as do zonae from oviductal ova. By usin a polyclonal antiserum raised against manually isolated zonae pellucidae of ov ductal eggs, Léveillé et al. (1987) also observed the addition of an oviductal ant gen to the zona. The reaction was most intense at the inner and outer surfaces the zona. The antigen(s) was also detected in the oviductal epithelium and flui within the bursa cavity. The lining of the uterine surface also showed a thin lay of fluorescent reaction. The latter observation probably explains why Fox and Shivers (1975) succeeded in detecting the oviductal antigen(s) on ova using absorbed anti-uterus sera.

Recently, monoclonal antibodies have been prepared against the oviductal glycoprotein associated with the zona pellucida (St-Jacques et al. 1986; Araki et al. 1986). These monoclonal antibodies are proving to be powerful tools to study the properties of this oviductal antigen.

Nomenclature

The oviduct secretes many different proteins, among which are those that are transferred to the egg. The latter proteins share the following properties:

1. Constitute polydispersed high molecular weight glycoproteins
2. Are synthesized by the oviductal secretory cells
3. Are secreted into the oviductal fluid
4. Become structural components of oviductal eggs and embryos

On the other hand, they appear to differ in several ways:

1. Localization differs according to the species considered; they may be found:
   a) Adhering to the surface of the zona pellucida as a thin granular deposit
   b) Deposited as a thick layer around the zona pellucida
   c) Adhering to the surface of the ooplasm
   d) Integrated within the thickness of the zona pellucida
2. Large variations in the molecular weights of the polymeric as well as monomeric forms
3. Differences in carbohydrate composition

It is certain that many other similarities and differences will be uncovered in the future. To obviate the need for a whole sentence in referring to these proteins, we propose the generic term "oviductin" for all proteins synthesized and secreted by the oviduct.

The glycoprotein secreted by the hamster oviduct and transferred to the zona pellucida would be such an oviductin.

When sufficient biochemical and immunological data have been collected to identify precisely and unequivocally a specific oviductal protein, it should bear the name oviductin followed by a letter.

Hamster Oviductin

Some preliminary information was recently obtained in our laboratory for the hamster oviductin using monoclonal antibodies against oviductal zonae pellucidae (unpublished):

- Native oviductin is of very high molecular weight (more than 400-kd). It is sensitive to protease digestion.

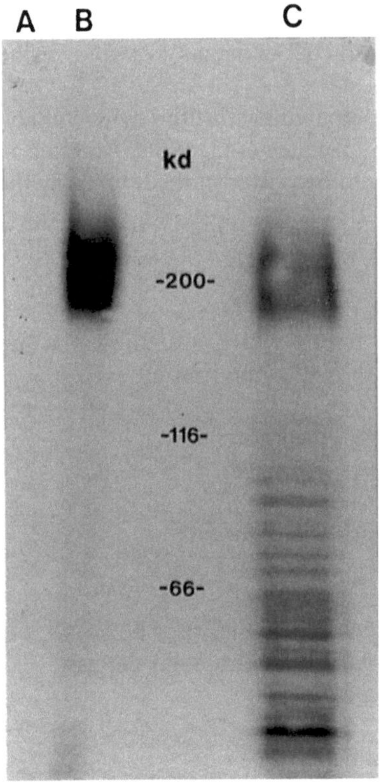

Fig. 1. Biosynthesis of oviductin by the hamster oviduct. Hamster oviducts were incubated witl [^{35}S]methionine. The labeled proteins secreted in the culture medium were analyzed by SDS PAGE under reducing conditions followed by electrotransfer on nitrocellulose and autoradiogra phy. Lane *C*, total ^{35}S-labeled secretory products; *lane B*, ^{35}S-labeled secretory products that wer immunoprecipitated using a monoclonal IgG against hamster oviductal zonae pellucidae; *lane A* control immunoprecipitation using a supernatant obtained from myeloma cell cultures. Oviducti in lane *B* appears as a polydispersed high molecular weight protein of approximately 200-kd (Robitaille et al., unpublished)

- Analysis by polyacrylamide gel electrophoresis (PAGE) under denaturating conditions (sodium dodecyl sulfate - dithiothreitol SDS-DTT) showed the monomeric form to be polydispersed with a molecular weight of approximately 200-kd.
- Tracer studies using [^{35}S]methionine confirmed that the oviductin is synthesized by the oviduct itself (Fig. 1).

Fig. 2. Localization of hamster oviductin at the electron-microscopic level using the protein A gold technique. Ultrathin sections of hamster oviduct (embedded in Lowicryl K4M) were incu bated with a monoclonal antibody (IgG) against hamster oviductal zona pellucida. This was fol lowed by incubation with protein A-gold and staining with uranyl acetate and lead citrate. Thi photomicrograph shows that the antigen (marked by the colloidal gold particles) is localize within the granules of the secretory cell. There was no specific labeling of the ciliated cell excep for the surface of the cilia in contact with the luminal content. (Kan et al., unpublished)

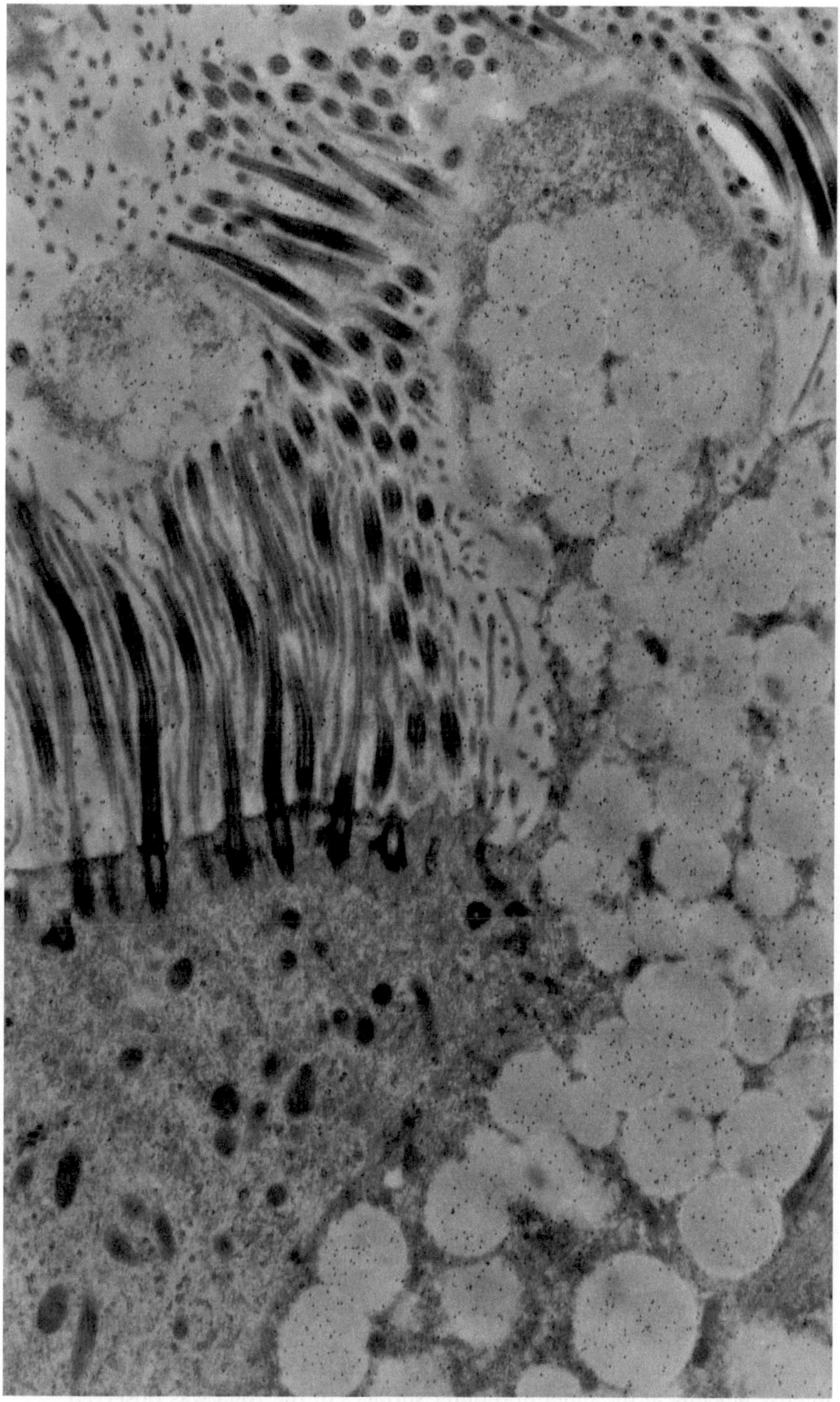

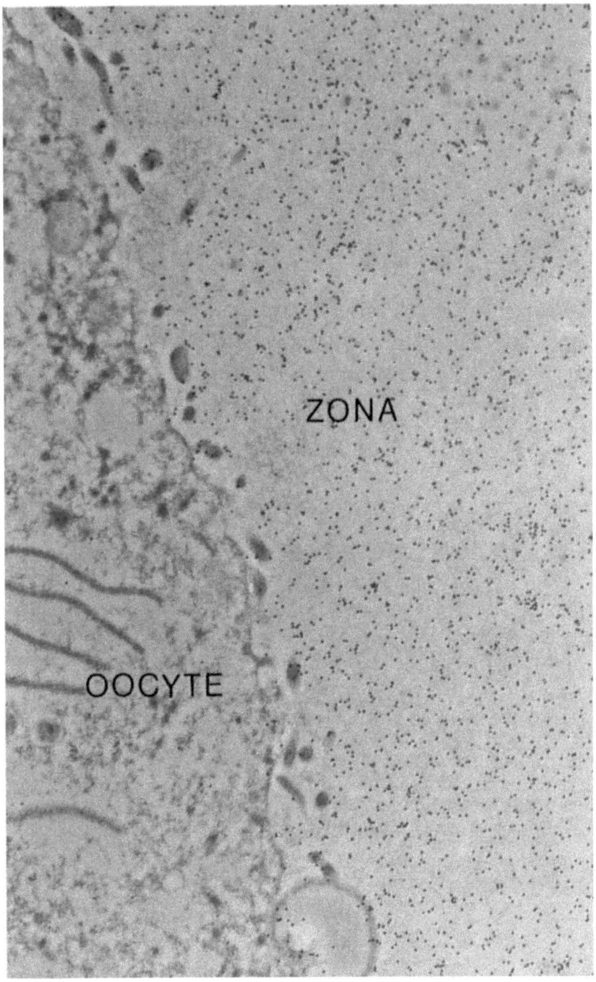

Fig. 3. Binding of hamster oviductin to the zona pellucida. Using the protein A-gold techniqu((see Fig. 2), it was observed that the antigen is localized evenly within the thickness of the zon; pellucida. The ooplasm as well as the plasma membrane of the oocyte were not labeled

- Treatment of native oviductin with trifluoromethanesulfonic acid showed tha this glycoprotein is composed of at least 70% carbohydrate. Molecular weigh determination of the protein core by radiation inactivation also confirmed th(very high carbohydrate content of the oviductin.
- In collaboration with Dr. F.W.K. Kan (Université de Montréal), it wa: observed, using the protein A-gold technique, that the hamster oviductin i: localized within the secretory cells of the oviduct and at the exterior surface o cilia (Fig. 2). It secondarily becomes evenly integrated within the thickness o the zona pellucida (Fig. 3). Using this immunocytochemical technique, n(intense labeling at the inner and outer surfaces of the zona was observed.

Biological Function of Oviductins

Discovery of the biological function(s) of oviductins will certainly be the major impetus to increase our knowledge of these proteins. Greenwald (1969) expressed "the disquieting possibility that oviductal secretions may be unessential, to a great extent, for embryonic development and merely represent changes in a secondary target organ to estrogen and progesterone." Even though Greenwald has attempted to disprove this possibility, it is only recently that significant improvement has been made in this field of reproduction.

Since the search for a biological function of oviductins is so important, the avenues of research most likely to be productive are presented here.

Oviductins could participate in the pickup and transport of the egg by the oviduct. This possibility stems from the work of Norwood et al. (1978) and Mahi-Brown and Yanagimachi (1983) who used chicken egg white as surrogate cumulus to facilitate these processes.

Lenz et al. (1982) made the observation that proteoglycans obtained from bovine follicular fluid can stimulate an acrosome reaction in bovine spermatozoa. It is not excluded that oviductins could also participate in different steps leading to fertilization.

Another potential biological function is that oviductins could react with and neutralize inhibitors of embryonic development described in rabbit oviductal fluid (Richardson et al. 1980) and human fetal cord serum (Ogawa and Marrs 1987).

By penetrating within the whole depth of the zona, oviductins could help to prevent the hardening of this structure. An effect of this type was recently reported for mouse oocytes (isolated from large antral follicles) exposed to fetal bovine serum (Downs et al. 1986).

Oviductins could serve as a host defence against adhesion of the zona (or even the hatched embryo) to the mucosal surface of the oviduct. The zona is coated with a glycoprotein (oviductin) that is present in the fluid in which it bathes as well as on the surface of ciliated and secretory cells. This makes it very unlikely that the zona can ever attach to the epithelium of the oviduct unless a pathological condition prevails. Only a thin film of oviduct-specific antigens was detected at the surface of the uterine wall (Léveillé et al. 1987). This might explain why the embryo can normally attach and begin implantation in the endometrium.

By concentrating in and/or on the zona, oviductins might protect the egg by dampening local changes in osmolarity or pH. High molecular weight glycoproteins (of the mucin type) are known to have this property. It is obvious that no steep osmotic or pH gradient could be overcome by such a barrier; it would more likely correspond to a fine-tuning mechanism. In this regard, it was reported that a small change of 5 mOsm in in vitro fertilization (IVF) culture media can influence embryonic development (Barg 1986).

The zona pellucida and the products of newly expressed embryonic and paternal genes can represent an immunologic challenge within the oviduct of the mother. Oviductins might prevent recognition of the zona surface by immunocompetent cells. Oviductins might also impede the passage of immunoglubulins targeted toward the embryo. This latter possibility might seem unlikely due to the fact that the zona was shown to be permeable not only to small molecular weight compounds (Austin and Lovelock 1958) but also to IgG and IgM at high concentrations (Hastings et al. 1972; Sellens and Jenkinson 1975) and even to virus (Gwatkin 1962). However, the level of immunoglobulins in oviductal fluid is much lower than in serum (Oliphant 1986). Sensitive immunocytochemical methods are now available which could be used to determine whether oviductins can protect the egg from natural antibodies (out of the pharmacological range).

The system of complement is another important component of the humoral immune system. The complement activity in tubal fluid is much less than the activity in the serum (Oliphant 1986). Oliphant et al. (1984a,b) have recently made the original observation that the sulfated glycoproteins that are synthesized and secreted by the oviductal epithelium can inhibit immunological complement from rabbit oviductal fluid. If this inhibiting material is shown to be associated with the egg, this would represent a most interesting biological function for oviductins as a local barrier against attack by the immune system.

Comment. A very high fertilization rate is the main mast of IVF clinics. However its sails, viable implantation have yet to catch their full wind. By completely bypassing oviductal functions we may very well be compromising the survival of the embryos. A better knowledge of the nature and biological function of oviductins might provide the solution to this problem.

References

Araki Y, Kurata SI, Oikawa T, Sendo F, Hiroi M (1986) Establishment of monoclonal antibody to a novel zona pellucida glycoprotein (ZP-O). 3rd International Congress of Reproductive Immunology, 1-5 July, Toronto

Austin CR, Lovelock JE (1958) Permeability of rabbit, rat and hamster egg membranes. Exp Cell Res 15: 260-61

Baranska W, Konwinski M, Kujawa M (1975) Fine structure of the zona pellucida of unfertilized egg cells and embryos. J Exp Zool 192: 193-202

Barg PE (1986) Small reduction of osmolarity of in vitro fertilization culture medium improves mouse embryo cleavage development (Abstr). Biol Reprod [Suppl 1] 34: 97

Barr SH, Oliphant G (1984) Sulfate incorporation into macromolecules produced by cultured oviductal epithelium. Biol Reprod 24: 952

Boland MP (1984) Use of the rabbit oviduct as a screening tool for the viability of mammalian eggs. Theriogenology 21: 127-137

Bourdage RJ, Halbert SA (1984) Quantitation of intraluminal motion of surrogate ova in the rabbit oviductal isthmus. Biol Reprod 30: 1124-1129

Braden AWH (1952) Properties of the membranes of rat and rabbit eggs. Aust J Sci Res [B] 5: 460

Brown CR, Cheng WTK (1985) Limited proteolysis of the porcine zona pellucida by homologous sperm acrosin. J Reprod Fertil 74: 257-260

Buhi WC (1986) Studies of synthesis and release of polypeptides by estrous and early pregnancy porcine oviduct tissue (Abstr). Int J Fertil 31: 392-393

Denker HW, Gerdes HJ (1979) The dynamic structure of rabbit blastocyst coverings. Anat Embryol (Berl) 157: 15-34

Downs SM, Schroeder AC, Eppig JJ (1986) Serum maintains the fertilizability of mouse oocytes matured in vitro by preventing hardening of the zona pellucida. Gamete Res 15: 115-122

Dunbar BS (1983) Antibodies to zona pellucida antigens and their role in fertility. In: Wegmann TG, Gill TJ III (eds) Immunology of reproduction. Oxford University Press, Oxford, pp 505-534

Erickson M, Ross P, Oliphant G (1986) Competitive ELISA quantitation of specific sulfated oviduct glycoproteins secreted from oviduct explants (Abstr). Biol Reprod [Suppl 1] 34: 74

Fazleabas AT, Verhage HG (1986) The detection of oviduct-specific proteins in the baboon (Papio anubis). Biol Reprod 35: 455-462
Fox L, Shivers CA (1975) Immunologic evidence for addition of oviductal components to the hamster zona pellucida. Fertil Steril 26: 599-608
Glass LE (1963) Transfer of native and foreign serum antigens to oviductal mouse eggs. Am Zool 3: 135-156
Greenwald GS (1969) Endocrinology of oviductal secretions. In: Hafez ESE, Blandau RJ (eds) The mammalian oviduct. University of Chicago Press, Chicago, pp 183-201
Gwatkin RBL (1962) Effect of virus on early mammalian development. I. Action of mengo encephalitis virus on mouse ova cultures in vitro. Proc Natl Acad Sci USA 50: 576-581
Hamner CE, Fox SB (1969) Biochemistry of oviductal secretions. In: Hafez ESE, Blandau RJ (eds) The mammalian oviduct. University of Chicago Press, Chicago, pp 333-355
Hastings RA II, Enders AC, Schlafke S (1972) Permeability of the zona pellucida to protein tracers. Biol Reprod 7: 288-296
Hirst PJ, de Mayo FJ, Dukelow WR (1981) Xenogenous fertilization of laboratory and domestic animals in the oviduct of the pseudopregnant rabbit. Theriogenology 15: 67-75
Kapur RP, Johnson LV (1985) An oviductal fluid glycoprotein associated with ovulated mouse ova and early embryos. Dev Biol 112: 89-93
Lawson RAS, Rowson LEA, Adams CE (1972) The development of cow eggs in the rabbit oviduct and their viability after re-transfer to heifers. J Reprod Fertil 28: 313-315
Lenz RW, Ax RL, Grimek HJ, First NI (1982) Proteoglycan from bovine follicular fluid stimulates an acrosome reaction in bovine spermatozoa. Biochem Biophys Res Commun 106: 1092-1098
Léveillé MC, Roberts KD, Chevalier S, Chapdelaine A, Bleau G (1987) Uptake of an oviductal antigen by the hamster zona pellucida. Biol Reprod 36: 227-238
Lippes J, van Oss CJ, Bronson PM, Alfonso LA, Dacalos E, Lucero R (1983) Human oviductal fluid proteins. II. Preparation of an antiserum to a human oviductal fluid protein: existence of autoantibodies against it in some sera. Fertil Steril 39: 824-828
Mahi-Brown CA, Yanagimachi R (1983) Parameters influencing ovum pickup by oviductal fimbria in the golden hamster. Gamete Res 8: 1-10
Mastroianni L Jr, Go KJ (1979) Tubal secretions. In: Beller FK, Schumacher GFB (eds) The biology of the fluids of the female genital tract. Elsevier/North-Holland, New York, pp 335-344
Mastroianni L, Urzua M, Stambaugh R (1970) Protein patterns in monkey oviductal fluid before and after ovulation. Fertil Steril 21: 817-820
Matsumoto H, Noda Y, Mori T (1986) Electrophoretic analysis of proteins in tissue extracts from endosalpinx and endometrium (Abstr). Int J Fertil 31: 401
Moghissi KS (1970) Human fallopian tube fluid. I. Protein composition. Fertil Steril 21: 821-829
Norwood JT, Hein CE, Halbert SA, Anderson RGW (1978) Polycationic macromolecules inhibit cilia-mediated ovum transport in the rabbit oviduct. Proc Natl Acad Sci USA 75: 4413-4416
Noske IG, Daniel JC (1974) Changes in oviduct fluid in golden hamster. J Reprod Fertil 38: 173-176
Ogawa T, Marrs RP (1987) The effect of protein supplementation on single-cell mouse embryos in vitro. Fertil Steril 47: 156-161
Oikawa T, Sendai Y (1984) Biochemical difference between zonae pellucidae of ovarian and oviductal eggs of the golden hamster (Abstr). Zool Sci 1: 932
Oliphant G (1986) Biochemistry and immunology of oviductal fluid. In: Seigler AM (ed) The fallopian tube: Basic studies and clinical contributions. Futura, Mount Kisco, pp 129-145
Oliphant G, Ross PR (1982) Demonstration of production and isolation of three sulfated glycoproteins from the rabbit oviduct. Biol Reprod 26: 537-544
Oliphant G, Cabot C, Ross P, Marta J (1984a) Control of the humoral immune system within the rabbit oviduct. Biol Reprod 31: 205-212
Oliphant G, Reynolds AB, Smith PF, Ross PR, Marta JS (1984b) Immunocytochemical localization and determination of hormone-induced synthesis of the sulfated oviductal glycoproteins. Biol Reprod 31: 165-174
Rao VH, Sarmah BC, Bhattacharyya NK (1984) Xenogenous fertilization of goat ova in the rabbit oviduct. J Reprod Fertil 71: 377-379

Richardson LL, Hamner CE, Oliphant G (1980) Some characteristics of an inhibitor of embryonic development from rabbit oviductal fluid. Biol Reprod 22: 553
Roberts GP, Parker JM, Symonds HW (1975) Proteins in the luminal fluid from the bovine oviduct. J Reprod Fertil 45: 301-313
Roberts GP, Parker JM, Symonds HW (1976) Macromolecular components of genital tract fluids from the sheep. J Reprod Fertil 48: 99-107
Sacco AG, Shivers CA (1973a) Effects of tissue-specific antisera on rabbit eggs. Biol Reprod 8: 481
Sacco AG, Shivers CA (1973b) Localization of tissue-specific antigens in the rabbit ovary, oviduct and uterus by the fluorescent antibody technique. J Reprod Fertil 32: 415-420
Sacco AG, Shivers CA (1973c) Comparison of antigens in the ovary, oviduct and uterus of the rabbit and other mammalian species. J Reprod Fertil 32: 421-427
Sellens MH, Jenkinson EJ (1975) Permeability of the mouse zona pellucida to immunoglobulin. J Reprod Fertil 42: 153-157
Shapiro SS, Jentsch JP, Yard AS (1971) Protein composition of rabbit oviducal fluid. J Reprod Fertil 24: 403-408
Shapiro SS, Brown NE, Yard AS (1974) Isolation of an acidic glycoprotein from rabbit oviductal fluid and its association with the egg coating. J Reprod Fertil 40: 281-290
Sirard MA, Lambert RD, Menard DP, Bedoya M (1985) Pregnancies after in vitro fertilization of cow follicular oocytes, their incubation in rabbit oviduct and their transfer to the cow uterus. J Reprod Fertil 75: 551-556
St-Jacques S, Forget A, Roberts KD, Chevalier S, Chapdelaine A, Bleau G (1986) Monoclonal antibodies specific for an oviductal component associated with the zona pellucida. 3rd International Congress of Reproductive Immunology, 1-5 July, Toronto
Stone SL, Richardson LL, Hamner CE, Oliphant G (1977) Partial characterisation of hormone-mediated inhibition of embryo development in rabbit oviduct fluid. Biol Reprod 16: 647-653
Sutton R, Nancarrow CD, Wallace ALC, Rigby NW (1984) Identification of an estrus-associated glycoprotein in oviductal fluid of the sheep. J Reprod Fertil 72: 415-422

Clinical Significance of Autoantibodies to the Zona Pellucida*

Takahide Mori, K. Ida, Y. Noda, H. Matsumoto, Tsuneatsu Mori, and E. Mori

Introduction

Recent studies have revealed that the zona pellucida possesses a significant biological role in reproductive processes and have shed light on a new aspect of human infertility. This chapter reviews the accumulation of clinical data in relation to the screening of antibodies to zona and discusses these data in terms of human infertility.

The zona pellucida is an extracellular glycoprotein matrix surrounding the mammalian ova from the stage of primordial follicle (Kang 1974) to the time of implantation of the blastcyst. It has recently been shown to be synthesized by the oocyte itself in mice (Bleil and Wassarman 1980; Bousquet et al. 1981) rather than by the ova or follicular cell, as formerly reported (Hedrick and Fry 1980). This envelope plays three major roles in fertilization and subsequent embryonic development: the recognition of homologous sperm, the blocking of polyspermy, and the compaction of blastomeres.

The heteroantibodies produced against zona material from rodents have been shown to interfere by passive immunization with fertilization in both in vitro studies (Oikawa and Yanagimachi 1975; Jilek and Pavlok 1975; Tsunoda and Chang 1976) and in vivo studies (Shahini et al. 1972; Oikawa and Yanagimachi 1975; Tsunoda and Chang 1976). The inhibitory effect of the antibody varies, depending upon the antigenic cross-reactivity between the two species - the immunogen source and the species immunized. It must be recalled that passive immunization with heteroantiserum to isolate zona inhibits sperm-egg interaction not only in the same species but also in other species (Tsunoda and Chang 1978). In vitro studies consistently show alloantibodies to be much weaker than heteroantibodies in interfering with fertilization both in active and in passive immunization (Tsunoda and Chang 1976); such results, however, are not as consistent in in vivo experiments. Mice are more susceptible to allogeneic zona antigens in producing alloantibodies than are rats. In hamster and rabbit, although the alloantibodies are detectable, they are not capable of blocking fertilization in vivo (Gwatkin et al. 1977; Wood et al. 1981).

These findings on experimental animals might tempt one to assume that the phenomenon is also common in humans. If alloantibodies - autoantibodies in this

* This work was supported, by a Grant-in-Aid for Scientific Research on Priority Area for the Ministry of Education (# 63640004).

case – exist in human sera, it would seem that they should be effective in interfer- ing with sperm penetration through zona.

In 1972 Ownby and Shivers reported that the antibody to zona pellucida blocked the penetration of sperm in hamster; this led to speculation that the phe- nomenon is an etiological factor in human infertility. In 1977 Shivers and Dunbar reported the existence of this antibody in the sera of infertile patients, as detected by indirect immunofluorescence (IIF) using porcine zona pellucida as target Examining the sera from 22 infertile patients, they found six patients with a strong and nine with a moderate immunofluorescent reaction. In 1978 we performed a similar IIF assay using porcine oocytes. Using absorption with formalin-fixed por- cine or human AB erythrocytes (Mori et al. 1978, 1979), we found the serum sam- ples of eight out of 52 infertile patients to be positive (15.4%) whereas the serum samples of only one of the 52 control subjects was positive (1.9%). In an IIF assay conducted in 1980 with a similar absorption procedure, the serum samples of 15 out of 297 infertile patients (5.1%) were positive (unexplained infertility in 13/175 or 7.4%) whereas none of the 141 control serum samples was positive (Nishimoto et al. 1980) (Table 1). Shivers made a comparative study using the IIF assay on anti-zona antibodies in the sera of clinically defined categories of patients regis tered at the World Health Organization (WHO) Reference Bank for Reproductive

Table 1. Indirect immunofluorescence assay

Infertile women	15/297 (5.1%)
Unexplained	13/175 (7.4%)
Anovulation	1/ 65 (1.5%)
Tubal and/or sperm factor	1/ 44 (2.3%)
Habitual abortion	0/ 13 (0%)
Controls	0/141 (0%)
Nonpregnant women	0/ 71 (0%)
Pregnant women	0/ 39 (0%)
Normal men	0/ 31 (0%)

(From Nishimoto et al. 1980)

Table 2. Zona reactivities in sera from WHO Reference Bank for Reproductive Immunology a detected by indirect immunofluorescence

Infertile women	7/43 (16.3%)
Unexplained	6/34 (17.6%)
Primary amenorrhea	0/ 2 (0%)
Recurrent abortions	1/ 7 (14.3%)
Control subjects	3/42 (7.1%)
Fertile women	0/14 (0%)
Pregnant women	2/ 9 (22.2%)
Postpartum women	0/14 (0%)
Virgins	1/ 4 (25.0%)
Infertile men	0/ 1 (0%)

(From Mori et al. 1985 a)

Immunology. As is shown in Table 2, the incidence of positive reactions was 16.3% in the infertile group compared to the 7.1% in the control group (Mori et al. 1985a). Trounson et al. (1980) reported similar findings obtained both by IIF and in in vitro fertilization studies.

However, Sacco and Moghissi (1979) demonstrated the same reactivity also in fertile women; as well as in men. Nayudu et al. (1982) reported that by a thorough absorption procedure the positive reaction in the IIF assay by the sera of infertile patients could be removed. Also Bleau et al. (1980) reported the absence of any positive fluorescence of infertile women on porcine zona when tested with human zona. A number of similar reports denying the existence of a human anti-zona antibody followed. The IIF test that has been widely used in detecting human anti-zona antisera is based upon well-established antigenic cross-reactivity between porcine and human zona pellucida (Shivers and Dunbar 1977; Sacco 1977; Takai et al. 1981; Dietl et al. 1982); this has been pointed out, however, to be somewhat subjective and to tend to be influenced by nonspecific binding. Therefore, use of this method alone may not be appropriate for elucidating the matter. New techniques have been developed which complement IIF in sensitivity and specificity. These include the passive hemagglutination reaction method (PHAR), radioimmunobinding assay (RIBA), and radioimmunoassay (RIA). These techniques are now applicable owing to the availability of the purified porcine zona substance in quantity.

In 1984 we examined the PHAR using bovine erythrocytes coated with solubilized zona material as indicator cells (Kamada et al. 1984). As is shown in Table 3, eight out of 88 (9.1%) infertile serum samples and only one sample from 90 control subjects (1.1%) showed positive reactions. Seven of the eight serum samples with positive PHAR results were demonstrated to be positive also in the IIF test. Absorption with fresh porcine erythrocytes caused positive reaction neither in PHAR nor in IIF on porcine oocytes. However, it is interesting that when absorbed with fresh human AB erythrocytes four out of eight serum samples with positive PHAR results retained the positive fluorescence on human zona. This fluorescence on human zona did not diminish by absorption with fresh porcine erythrocytes.

These findings are in agreement with the reports by Dakhno et al. (1980) and

Table 3. Passive hemagglutination reaction

Infertile women	8/88 (9.1%)
Unexplained	3/23 (13.0%)
Anovulation	4/35 (11.4%)
Tubal obstruction	0/ 9 (0%)
Sperm abnormality	1/14 (7.1%)
Uterine abnormality	0/ 7 (0%)
Controls	1/90 (1.1%)
Nonpregnant women	1/59 (1.7%)
Pregnant women	0/18 (0%)
Normal men	0/13 (0%)

(From Kamada et al. 1984)

Table 4. Anti-zona activities in sera from WHO Reference Bank of Reproductive Immunology as detected by passive hemagglutination reaction[a]

Infertile women	3/96	(3.1%)
Unexplained	2/83 (1:8, 1:16)	(2.4%)
Primary amenorrhea	1/ 2 (1:32)	(50.0%)
Recurrent abortions	0/10	(0%)
Autoimmune ovarian failure	0/ 1	(0%)
Control subjects	2/222	(0.9%)
Women		
Fertile	0/22	(0%)
Pregnant	1/35 (1:16)	(2.9%)
Postpartum	1/16 (1:16)	(6.3%)
Postmenopausal	0/ 7	(0%)
Virgin	0/12	(0%)
Men		
Fertile	0/26	(0%)
Unexplained sterility	0/81	(0%)
Vasectomized	0/15	(0%)
Congenital absence of vas	0/ 3	(0%)
Aspermatogenesis	0/ 5	(0%)

(From Mori et al. 1985a)

[a] Antibody titers are shown in parentheses.

Nayudu et al. (1982) who also showed the higher incidence of positive PHAR results in sera of infertile patients than in that of control subjects. Tests by PHAR on a total of 318 samples from the WHO Reference Bank for Reproductive Immunology (Mori et al. 1985a) revealed a 3.1% positivity (3/96) among infertile women compared to 0.9% (2/222) among control subjects (Table 4). There were no discrepancies in the results between two identical samples included in duplicate. With a different method of PHAR, using human 0 Rh(+) erythrocyte as indicator cells, Dunbar tested 297 samples of WHO sera and found no positive reaction (Mori et al. 1985a).

Using the WHO samples and the RIBA methods Czuppon et al. (1987) obtained a 14.3% positive reaction in the unexplained infertile group compared to 0% in control women, but 19.6% in control men. Surprisingly, a high incidence of positive reaction was seen in vasectomized men (46.2%); it was speculated that this may be due to cross-reactivity between spermatozoal component and zona material (Mori et al. 1985a).

Numerous clinical investigations have been carried out using various methods, and those cited above report only a few of these. Findings, however, have not always been consistent. We therefore considered it necessary to ascertain whether the anti-zona antibody detected in human sera does in fact hinder fertilization in vitro and may thus be the causal factor of human infertility. The development of reliable methods both in screening of the antibody and in the biological analysis of the penetration of sperm has been awaited. In this line of study we used ^{125}I-labeled protein A solid-phase RIA to screen the antibody and performed further analysis using the zona penetration test on highly concentrated salt-stored human oocytes.

Materials and Methods

Preparation of Zona Material

Frozen pig ovaries were purchased at a local slaughter house, kept at −20 °C, and thawed just prior to use. Porcine ovarian follicles were disrupted by a specially devised machine: two multineedled disks were set facing each other so as to stick to the ovaries and disrupt them when the upper disk turned, while the other remained still. Disrupted ovaries were washed with phosphate-buffer saline (PBS); recovered PBS was filtered first by a sheet of 1-mm metal net and then by a sheet of 250-μm Saran mesh to remove the tissue debris. Recovered PBS containing oocytes was further cleaned of debris by a rigorous glass-wool treatment. All these procedures were performed under refrigeration to avoid degeneration of the material. The oocytes were trapped finally in a 100-μm salan mesh. These were suspended in 20 ml PBS, layered on 30 ml 50% sucrose solution, and centrifuged at 1500 *g* for 30 min. From the interphase between PBS and sucrose oocytes with intact zona were recovered (Fig. 1 A). These were gently homogenized, and the homogenate was again centrifuged on 50% sucrose at 1500 *g* for 30 min. Oocyte-free zona skeletons were collected from the precipitate (Fig. 1 B).

These were dialyzed against distilled water, lyophilized, and solubilized in 0.1-*M* sodium borate buffer at 60 °C, ph 9.0, for 2 h or stored at −20 °C (Noda et al. 1981)

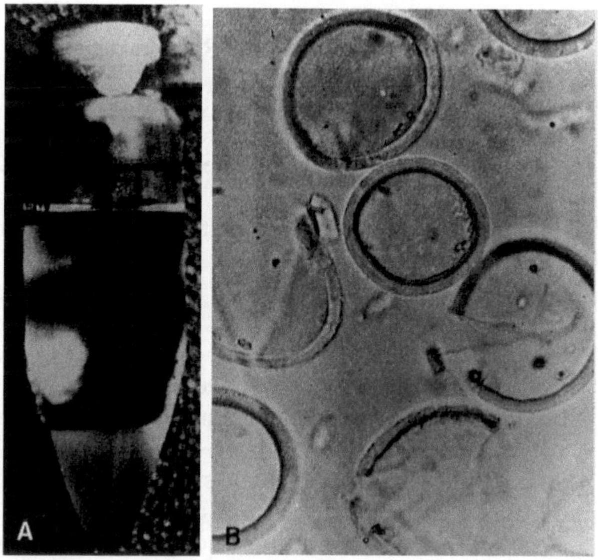

Fig. 1 A, B. Isolation of porcine zona pellucida. **A** Centrifugation in 50% sucrose. Porcine oocytes are seen to be trapped in interphase. **B** Recovered zona pellucida

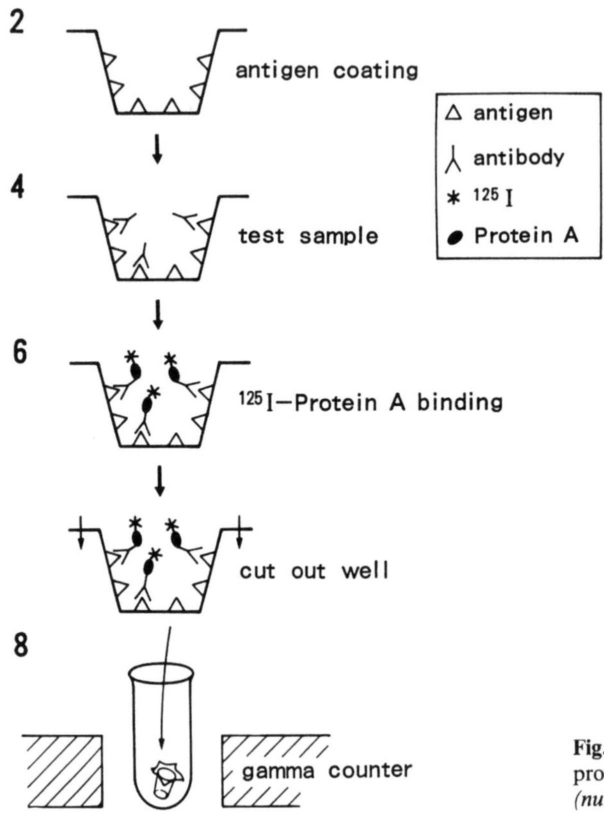

Fig. 2. ^{125}I-labeled protein A RIA procedure. For explanation of steps *(numerals)* see text

^{125}I-Labeled Protein A Solid-Phase Radioimmunoassay

Figure 2 shows the method of ^{125}I-labeled protein A RIA along with the number of procedures. The serum samples collected were inactivated by heating at 56 °C for 30 min, divided into several tubes, and stored at −20 °C until use. Just before the experiment the serum samples were thawed and centrifuged at 20000 *g* for 20 min to exclude immunoprecipitations. The procedure followed was as follows:

1. 50 µl zona solution in 0.1-*M* PBS was poured into the wells of a flexible plastic micro titer plate.
2. The zona antigen was dried and immobilized to the flat bottom of individual wells in an incubator at 37 °C overnight.
3. 100 µl 5% bovine serum albumin (BSA) in PBS was added to each well and incubated at 37 °C for 1 h.
4. After decantation 50 µl immune serum in 1/4 dilution with 1% PBS containing 5% BSA was added to each well and incubated for 12 h at 4 °C.
5. Each well was washed with PBS containing 0.01% NP40, a detergent.
6. Fifty µl ^{125}I-labeled protein A in 0.1% PBS, adjusted to 6000 cpm/ng with cold protein A, was added to the well and incubated at 4 °C overnight.

7. After being thoroughly washed with PBS each well was cut out.
8. The well cut out was placed in a 12 × 75 mm plastic test tube and measured in a gamma counter.

Indirect Immunofluoresence Assay

Porcine oocytes were collected either with follicular aspiration or dissection of ovaries. Cumulus cells were removed mechanically by pipetting the oocytes several times with pipets having a bore size slightly smaller than the oocyte plus its zona and were thoroughly washed with PBS. The oocytes were incubated at 20 °C for 2 h with the patient's serum in 1/4 dilution and were washed first with forceful pipetting and then laid in PBS for 2 h. They were then incubated with goat anti-human IgG serum labeled with fluorescein isothiocyanate at 1/20 dilution for 10–30 min and were washed as described above. Each test was carried out in triplicate with oocytes from the same donor.

Zona Penetration Test

Human oocytes obtained at laparotomy from the ovaries of patients with severe gynecological disease (with their informed consent) were stored in a highly concentrated salt solution of 0.5-*M* $(NH_4)_2SO_4$, 1. 0-*M* $MgCl_2$, 0.1% dextran at 4 °C (Yanagimachi et al. 1979) for up to 4 months. The oocytes were desalted with BWW medium overnight before use. They were then incubated with each sera for 1 h at 37 °C under 5% CO_2 in air. After incubation the oocytes were tested for sperm penetration. For insemination freshly ejaculated spermatozoa from a healthy male with consistently high sperm quality were washed twice with BWW with control serum of 7.5% by centrifugation at 220 *g* for 5 min. The sperm pellet was gently overlaid with 2 ml BWW medium in the tube which was set at a 30 ° angle in a CO_2 incubator for 1 h to let the sperm swim up into the medium. The suspension of motile spermatozoa was recovered and added to the oocytes in the medium covered by paraffin oil in a plastic dish with the sperm concentration of $0.5\text{–}1.0 \times 10^6$/ml. After the incubation at 37 °C under 5% CO_2 in air for 6 h oocytes were examined under a phase-contrast microscope for sperm penetration through the zona pellucida.

Monoclonal Antibodies

Two 5-week-old female mice (BALB/C) were immunized by subcutaneous injection with a mixture of 0.25 ml complete Freund's adjuvant and 0.25 ml solubilized zona pellucida antigen at a protein concentration of 30 µg/ml prepared as described above. After 4 weeks the mice were given an intramuscular injection of 0.25 ml of the same zona antigen, and 3 days later their spleen was removed. The mouse spleen cells (1.5×10^8) were fused with mouse myeloma cells (2.4×10^7) in the presence of 50% polyethylene glycol (PEG 3000). The cells were then distributed into tissue culture microplates and hybrid cells were selected in hypoxanthine aminopterine thymidine (HAT) medium (Roswell Park Memorial Institute, RPMI, 1640 containing 15% fetal calf serum, 2 m*M* L-glutamine, 1 m*M* pyruvate. 0.1 mg/

Table 5. ^{125}I-Labeled Protein A solid-phase radioimmunoassay

	Mean (cpm × 10^3)	No. of Samples
Infertile patients	25.6 ± 3.7	(95)
Unexplained	25.1 ± 4.4	(39)
Anovulatory	27.1 ± 5.7	(13)
Tubal obstruction	27.0 ± 5.6	(11)
Sperm abnormality	24.8 ± 4.7	(27)
Uterine abnormality	28.0 ± 4.9	(5)
Autoimmune disease patients	27.3 ± 5.5	(33)
Controls	22.5 ± 3.6	(24)
Fertile women	22.6 ± 3.6	(14)
Normal men	22.4 ± 3.5	(10)

ml kanamycin). Each of the hybridomas producing anti-zona antibodies was cloned by the limiting dilution method. The limiting dilution was performed three times, three hybridoma cells per well once and one cell per well twice.

Results and Discussion

^{125}I-Labeled Protein A RIA

A total of 152 serum samples were examined. Samples were from 95 infertile patients, 37 of them being unexplained cases, 33 autoimmune disease patients, and as control 14 fertile women and 10 men. The average cpm titers (× 10^3) of these groups were 25.6 ± 3.7 for infertile patients, 27.3 ± 5.5 for those with autoimmune disease, and 22.6 ± 3.6 and 22.4 ± 3.5 for fertile women and men, respectively (Table 5). The cpm titer of the samples was plotted in a chart to show the diversity of the count in each group (Fig. 3). The average cpm titer of the unexplained infertility patients (25.1 ± 4.4) was not higher than that of the infertility group on the whole, nor was the distribution pattern divergent. However, the distribution of autoimmune disease patients, mostly systemic lupus erythematodes (SLE) patients, showed the most dispersed pattern, with the average cpm titer being the highest of all the groups. The highest titer was also recorded in this group (42.1 × 10^3 cpm). Fertile women and men (control groups) showed a similar pattern, with a slightly low average and somewhat less divergent pattern. None of these control samples exhibited high responses.

Seven cases were selected from those which showed high titers, four from the infertile group and three from the autoimmune disease group; these were numbered H_1 to H_7. Similarly, samples from six subjects in the low-titer group were selected; two infertile patients, two autoimmune disease patients, one fertile women, and one fertile man; these were numbered L_1 to L_6. (The 13 cases are marked with squares in Fig. 3.) These serum samples were further investigated with the IIF and zona penetration test. In the high-titer group, among the fou

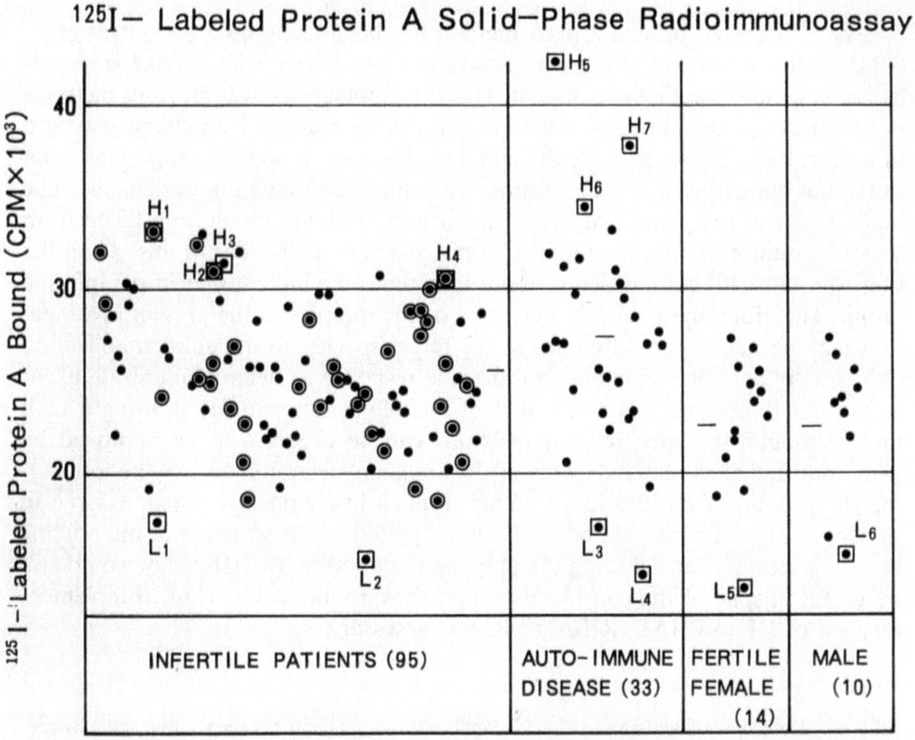

Fig. 3. Anti-zona activity of serum samples as detected by ^{125}I-labeled protein A radioimmunoassay. *Dashes,* average titer count for each group; *squares,* samples analyzed by immunofluorescence and zona penetration test; *filled circles, squares,* unexplained infertility

infertile patients selected two fell in the diagnostic category of unexplained infertility (H_1, H_4). The other two were cases of oligospermy (H_3) and of tubal obstruction (H_2). The latter had previously possessed antisperm antibodies, as detected by the sperm immobilization test, but the recent examination revealed a decrease in the titer. Interestingly, this patient conceived 8 months after this investigation. However, our preliminary data showed no correlation between the sperm immobilizing test and anti-zona activity.

The three samples selected from the 26 patients with autoimmune disease were those which exhibited the highest titers of all the samples assayed. The finding that patients with autoimmune disease show a high titer of anti-zona antibody is in line with the findings obtained on thymectomized mouse. One of the three patients (H_5) had been affected with SLE for 12 years. Of interest is the fact that in these clinical activity of the autoimmune disease showed a parallel correlation to the anti-zona activity. In two out of three cases of SLE zona activity as examined by the RIA count was seen to rise in the active state of SLE and fall in the inactive state. Considered together with the conceived case of H_2, this may indicate the possibility of fluctuation in the titer within the patient herself according to the immunological state at that moment.

The ^{125}I-labeled protein A RIA method has been lately used by Gulyas et al. (1983) in the titration of monkey anti-zona sera immunized with porcine zona pellucida in a contraception experiment. This is advantageous over previous methods in that the large quantities of collected sera can be examined simultaneously and in a short time. However, since the target of this assay is solubilized porcine zona instead of human zona, a high response by human sera cannot be anticipated. The highest titer in the preliminary experiment was exhibited by the sera taken from heteroimmunized rabbit immunized by porcine zona pellucida. In this serum the titer was 150×10^3 cpm, which is about four times the highest titer in the infertile group. Therefore, the relatively low titers of the samples in the present investigation may be due to the difference in the target species, to the autoimmunization (which may not elicit as intense an immune response as heteroimmunization), or to the possibility that this assay did not include serum samples showing a high titer. Although the specificity of reactions can be expected to be improved by using purified zona material, some of the antigenic determinants may be lost during the process of purification, leading to much lower positive reactions even in patients with anti-zona antibodies. If an integrated whole structure of the porcine zona is necessary for detecting the anti-zona antibodies, the IIF assay would still be useful in spite of the considerably high background levels. Thus, complementary use of IIF to PHAR, RIBA, or RIA is necessary.

Table 6. Detection of anti-zona pellucida antibody: comparison between high- and low-titer groups of ^{125}I-labeled protein A assay

Sample subjects	^{125}I-Protein A RIA cpm($\times 10^3$)	IIF	Zona penetration test (No. of sperm penetrated)				
			1st experiment		2nd experiment		Penetration%
High-titer group							(2/18)
Infertile							11.1%
H_1 (300)	32.3	++	–	–	0	0	
H_2 (318)	31.2	++	0	–	0	0	
H_3 (320)	31.6	++	–	–	0	0	
H_4 (382)	31.7	+	0	–	–	–	
Autoimmune							
H_5 (972)	42.1	+	0	0	0	0	
H_6 (2007)	34.2	++	–	–	1	0	
H_7 (2035)	37.0	++	0	0	1	0	
Low-titer group							(8/11)
Infertile							72.8%
L_1 (301)	17.6	–	–	–	1	0	
L_2 (360)	16.4	–	–	–	6	1	
Autoimmune							
L_3 (2021)	17.1	–	–	–	0	–	
L_4 (2039)	14.3	–	–	–	3	–	
Fertile woman							
L_5 (F-8)	13.9	±	3	2	–	–	
Fertile man							
L_6 (M-6)	19.5	–	1	0	1	–	

RIA, radioimmunoassay; IIF, indirect immunofluoresence assay

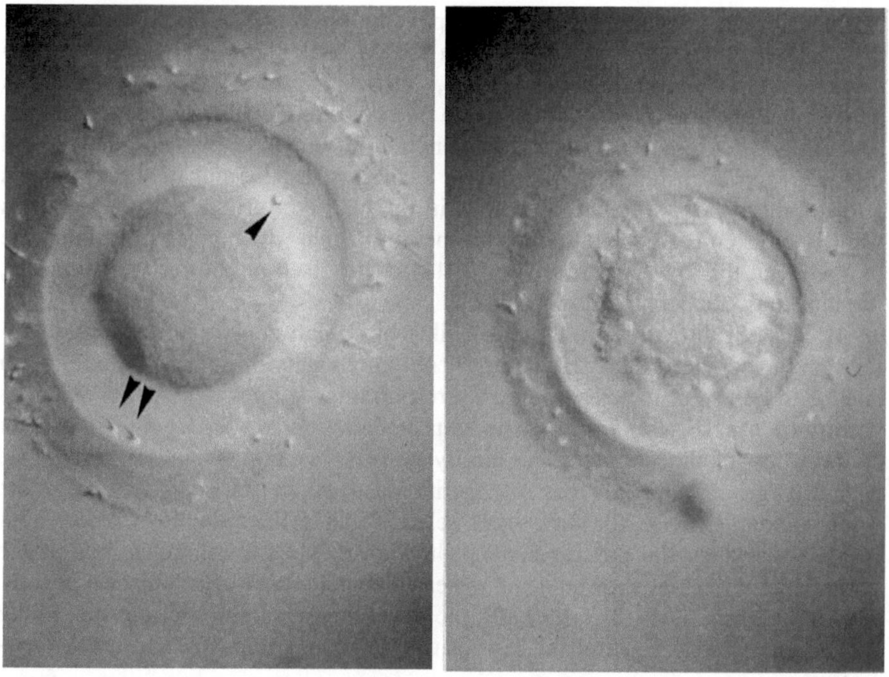

Fig. 4. Zona penetration test. Phase contrast micrograph of human oocyte stored in highly concentrated salt salution. *Left,* penetrated by three sperm; *right,* not penetrated

Indirect Immunofluorescence Test

The 13 sera from the H and L groups were examined by IIF with porcine oocytes as target. As shown in Table 6, all seven samples in the high-titer group produced positive fluorescence (five strong and two moderate). Five out of six in the low-titer group did not; the one remaining was of a fertile female and showed slightly positive fluorescence. The fluorescence produced on the porcine oocytes were all located at the outer rim of the zona. The absorption was done using porcine oocytes, resulting in the removal of all the positive fluorescence by IIF.

Zona Penetration Test

The 13 serum samples from the H and L groups were also examined by the zona penetration test (Fig. 4). This method, devised by Yanagimachi et. al. (1979) and utilizing highly concentrated salt-stored human oocytes, has been proved efficient by Tsukui et al. (1986) in the investigation of antisperm antibody.

Two experiments were done using the zona penetration test with human oocytes as target; ten oocytes in the first experiment and 19 in the second experiment. In the first experiment all four sera (H_2, H_4, H_5, H_7) of the high-titer group blocked the penetration of the sperm through the zona in every oocyte used, whereas in the low-titer group one serum (H_5) allowed the penetration in both of two oocytes, and the other (H_6) allowed penetration in one of the two oocytes. In the second experiment four samples in the high-titer group (H_1, H_2, H_3, H_5) blocked the penetration in both oocytes, and two samples (H_6, H_7) blocked the penetration in one. In the low-titer group one sample (L_2) allowed the penetration on both oocytes, one (L_1) allowed the penetration on one of the two oocytes, and in three samples using only one oocyte each two (L_4, L_6) out of three samples allowed the penetration. On the whole, in the high-titer group, out of 18 oocytes used 16 blocked the penetration of the sperm to zona, which gives a blocking rate of 88.9%; in the low-titer group, out of 11 oocytes used only three were blocked, the blocking rate being 27.2% (Table 6).

Because of difficulties in the availability of human oocytes not enough investigation has yet been done. However, these experiments show ^{125}I-labeled protein A RIA to be useful in screening highly responding samples to zona pellucida, as confirmed by IIF test, and that the zona penetration test is effective in further examination of the samples. The mostly positive reaction to the IIF test in the high-titer group and the mostly negative reaction in the low-titer group are understandable, since the sera were selected beforehand to be responsive or nonresponsive by the porcine zona pellucida itself. Furthermore, results of the zona penetration test showed a quite clear difference between these two groups: most of the sera of the high-titer group blocked the penetration of the sperm into the oocyte whereas most of the sera of the low-titer group failed to block penetration.

Monoclonal Antibodies

The question arises as to whether the same result would be obtained with anti-zona monoclonal antibodies. We therefore produced monoclonal antibodies against porcine zona material for further investigation (Mori et. al. 1985b). Two monoclonal antibodies to antigens common to the zona pellucida of pigs and hamsters were obtained. At high concentrations these two monoclonal antibodies prevented sperm binding and penetration into human zona pellucida in vitro. However, a monoclonal antibody directed against an antigen restricted to porcine zona pellucida did not have these inhibitory effects.

Hybridoma cell lines showing high antibody titers to the zona pellucida antigen were selected by the limiting dilution and with the PHAR test as described above. The PHAR titers of the supernatants from six hybridoma cell lines obtained ranged from 1:60 to 1:5120 and were not reduced by absorption with any tissues or erythrocytes tested. On immunodiffusion, supernatants of each of these cell lines formed a precipitation line with rabbit anti-mouse IgG1, but not with rabbit anti-mouse IgG2α, G2β, or G3 antiserum or with goat anti-mouse IgM antiserum, indicating that all six monoclonal antibodies belonged to the IgG1 subclass.

The IIF test was performed with these monoclonal antibodies. The supernatants of all six hybridoma clones stained the entire layer of porcine zona pellucida. Three of the supernatants gave positive reactions against human oocytes. One of the supernatants (number 2) stained the entire zona, the second (number 5) stained only the outer layer, and the last stained the inner layer or the vitelline

Table 7. Zona penetration test on human oocytes with monoclonal antibodies to porcine zona pellucida

Monoclonal antibodies	Antibody titer by PHA test	Experiment 1 No. of sperm		Experiment 2 No. of sperm		Experiment 3 No. of sperm	
		Bound	Penetrated	Bound	Penetrated	Bound	Penetrated
No. 2	160	0	0	0	0	0	0
	8	16	17	n.d.	n.d.	n.d.	n.d.
No. 5	160	0	0	0	0	0	0
	8	17	12	n.d.	n.d.	n.d.	n.d.
Culture medium	0	9	15	n.d.	n.d.	n.d.	n.d.
Control serum	0	11	4	31	30	9	3

n.d., not done

membrane. No monoclonal antibodies gave a positive reaction in the IIF tests against mouse, rat, or hamster oocytes.

Monoclonal antibodies numbers 2 and 5 were further used in the zona penetration test. Three experiments were performed with human oocytes from different donors. As summarized in Table 7, these two monoclonal antibodies at a titer of 160 PHAR prevented both the binding and the penetration of sperm. When the monoclonal dilutions were lowered to a titer of 8 PHAR, no inhibitory effects were observed. Neither sperm binding nor penetration was inhibited by pretreatment of oocytes with monoclonal antibody number 1, which recognized porcine but not human zona pellucida in the IIF test. These results are in essential agreement with those reported by Isojima et al. (1984), although our monoclonal antibodies differed from theirs in that ours showed different staining patterns in the IIF tests on oocytes in various species.

Thus, we have seen that not only the polyclonal antibodies in human sera but also monoclonal antibodies in mice interfere with the penetration of sperm through the zona of corresponding oocytes. We also noticed that some of the autoimmune patients possessed a considerable amount of anti-zona activity as detected by ^{125}I-labeled protein A RIA.

Thymectomized Mice

Could there be an animal model for the biological phenomenon of auto-anti-zona antibody? Working on thymectomized mice we have recently obtained interesting results. Autoantibodies to the zona are known to be induced experimentally in mice by neonatal thymectomy. Most of the mice thymectomized on the 3rd day after birth are seized with oophoritis. This type of oophoritis is characterized by the appearance of autoantibodies to zona, cytoplasma, and vitelline membrane of ova and steroid-producing cells of granulosa and theca cells, leading to ovarian dysgenesis and subsequent sterility after puberty (Nishizuka and Sakakura 1969, 1971). Immune tolerance against ovarian autoantigens is sustained by helper and suppressor T cells, both of which are specifically sensitized to ovarian antigen at a

Table 8. Correlation between titers of anti-zona pellucida antibodies and fertilization-blocking effects

Serum		Titer of anti-zona pellucida antibodies							
		1	2	4	8	16	32	64	128
Anti-zona serum No. 30									
	Fertilization rate (%)	–	–	–	78.6[b]	60.0[a]	28.1[a]	3.6[a]	0.0[a]
	Light scattering								
	Fluorescent	+	+	+	+	++	++	+++	+++
Anti-zona serum No. 6									
	Fertilization rate (%)	75.0[b]	80.8[b]	79.2[b]	89.3[b]	–	–	–	–
	Light scattering					–	–	–	–
	Fluorescent	+	+	+	+	–	–	–	–
Control	Fertilization rate (%)	75.0	78.0	78.8	82.2	90.0	76.7	78.4	87.0

χ^2 test against control: [a] $p < 0.001$; [b] not significant

certain time after birth (Nishizuka 1982). Anti-zona autoantibodies are detected in 10%–20% of these animals. We examined one serum (number 30) against oocytes from normal mouse. The intensity of immunofluorescence and light scattering increased as the titer of the antizona reaction, determined by the PHAR method, increased (Table 8). The photomicrograph of the results of IIF with this serum shows positive fluorescence together with negative control sera (Fig. 5). The photomicrograph of zona scattering with this serum is shown with control sera in Fig. 6. When we tested the serum in a fertilization experiment, as the titer increased, fertilization decreased, reaching a complete block at the PHAR titer of 128. These data thus indicate that the neonatal thymectomized mice produced auto-anti-zona antibodies which blocked the penetration of sperm through the zona pellucida in vitro. Neonatal thymectomy may cause an imbalance of the T-cell population with enhanced helper and/or reduced suppressor functions; this results in subsequent derangement of immune tolerance against ovarian antigens, leading in turn to the production of autoantibodies to these antigens (Taguchi and Nishizuka 1980; Taguchi et al. 1980). Tung et al. (1987) also reported the inhibition of fertilization in vitro. Their finding that serum anti-oocytes/zona antibody subsided as inflammation in the ovaries subsided is interesting in relation to our clinical findings that the activity of the autoimmune disease coincides with the anti-zona activity in some SLE patients.

Although further investigation is required to determine whether the thymectomized autoantibody in mice is identical to the spontaneous autoantibodies against zona pellucida in humans, the data presented here implies the possibility of the existence and an inhibitory action of these antibodies on human fertilization.

A considerable amount of data have been accumulated in this field. The pathophysiological significance of this anti-zona antibody as a possible cause of infertil

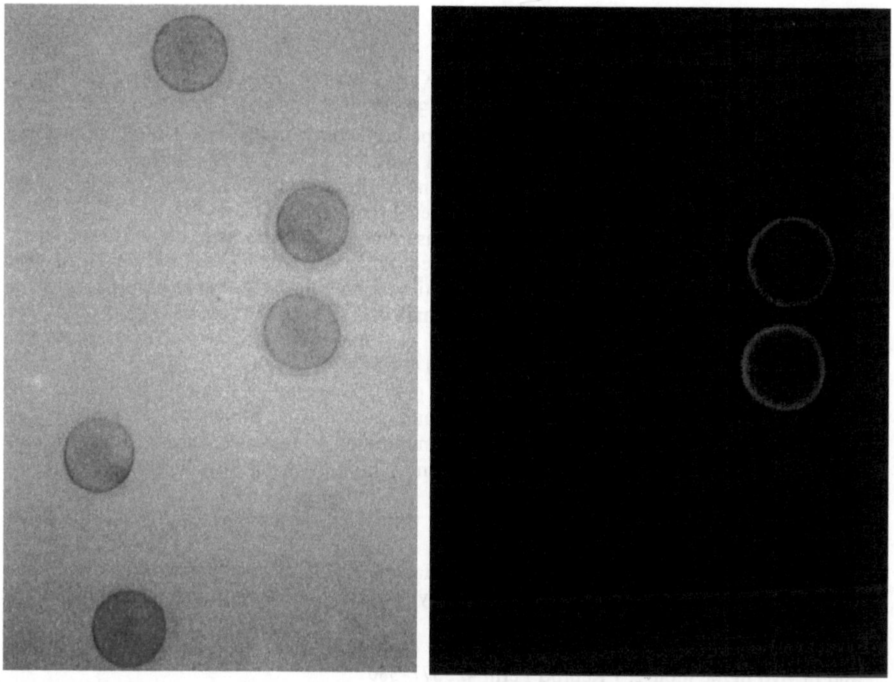

Fig. 5. Micrograph of indirect immunofluorescence. Two out of five mouse oocytes incubated with auto-anti-zona antibody positive serum of thymectomized mice show positive fluorescence, whereas the other three, incubated with control sera, seen on phase contrast micrograph have no fluorescence reaction

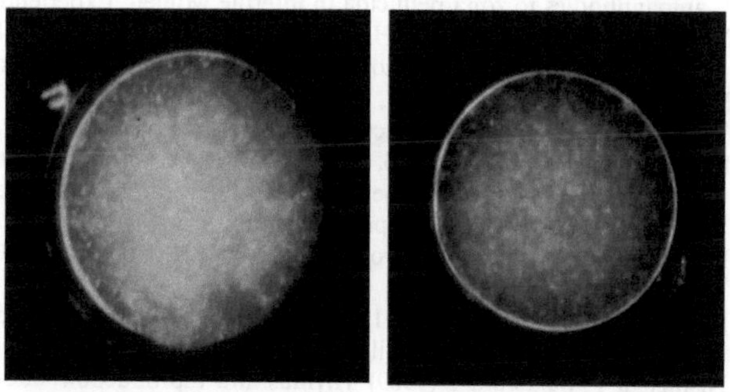

Fig. 6. Formation of immunoprecipitate on mouse oocytes by auto-anti-zona antibody produced by the thymectomized mice. *Left,* positive; *right,* negative

ity in women has not yet been clearly elucidated, and there remain various controversies and problems yet to be solved. However, we would like to conclude this chapter by stressing the possibility of this antibody being one of the causes of human infertility.

References

Bleau G, Bousquet D, Roberts KD, Chapdelaine A, Faucher G (1980) Anti-gamete antibodies in both partners of an infertile couple. Fertil Steril 34: 198

Bleil JD, Wassarman PM (1980) Synthesis of zona pellucida proteins by denuded follicle-enclosed mouse oocytes during culture in vitro. Proc Natl Acad Sci USA 77: 1029-1033

Bousquet D, Leveille MC, Roberts KD, Chapdelaine A, Bleau G (1981) The cellular origin of the zona pellucida antigen in the human and hamster. J Exp Zool 215: 215-218

Czuppon AB, Dietl J, Tripatzis J (1987) Solid-phase antigen immunoassay for the detection of anti-zona pellucida-antibodies in clinically defined sera. J Clin Chem Clin Biochem 25: 87-90

Dakhno FV, Hijort T, Grishenko VI (1980) Evaluation of immunofluorescence on pig zona pellucida for detection of anti-zona antibodies in human sera. J Reprod Immunol 2: 281-291

Dietl J, Knop G, Mettler L (1982) The frequency of serological anti-zona pellucida activity in males, females and children. J Reprod Immunol 4: 123-131

Gulyas BJ, Yuan LC, Gwatkin RBL, Schmell ED (1983) Response of monkeys to porcine zona pellucida as detected by a solid-phase radioimmunoassay. J Med Primatol 12: 331-342

Gwatkin RBL, Williams DT, Carlo DJ (1977) Immunization of mice with heat-solubilized hamster zonae; production of anti-zona antibody and inhibition of fertility. Fertil Steril 28: 871-877

Hedrick JL, Fry GN (1980) Immunocytochemical studies on the porcine zona pellucida. J Cell Biol 87: 1025

Isojima S, Koyama K, Hasegawa A, Tsunoda Y, Hanada A (1984) Monoclonal antibodies to porcine zona pellucida antigens and their inhibitory effects on fertilization. J Reprod Immunol 6: 77-87

Jilek F, Pavlok A (1975) Antibodies against mouse ovaries and their effect on fertilization in vitro and in vivo in the mouse. J Reprod Fertil 42: 377-380

Kamada M, Hasebe H, Irahara M, Kinoshita T, Naka O, Mori T (1984) Detection of anti-zona pellucida activities in human sera by the passive hemagglutination reaction. Fertil Steril 41: 901-906

Kang YH (1974) Development of the zona pellucida in the rat oocyte. Am J Anat 139: 535-566

Mori T, Nishimoto T, Kitagawa M, Noda Y, Nishimura T, Oikawa T (1978) Possible presence of autoantibodies to zona pellucida in infertile women. Experientia 34: 797-799

Mori T, Nishimoto T, Kohda H, Takai I, Nishimura T, Oikawa T (1979) A method for specific detection of autoantibodies to the zona pellucida in infertile women. Fertil Steril 32: 67-72

Mori T, Kamada M, Hasebe H, Irahara M, Shivers CA, Czuppon AB, Mettler L, Mathur S, Dunbar BS (1985 a) Antibody reactivity with porcine zona pellucida. J Reprod Immunol 8: 337-345

Mori T, Kamada M, Yamano S, Kinoshita T, Kano K, Mori T (1985 b) Production of monoclonal antibodies to porcine zona pellucida and their inhibition of sperm penetration through human zona pellucida in vitro. J Reprod Immunol 8: 1-11

Nayudu PL, Freemann LE, Trounson AO (1982) Zona pellucida antibodies in human sera. J Reprod Fertil 65: 77-84

Nishimoto T, Mori T, Yamada I, Nishimura T (1980) Autoantibodies to zona pellucida in infertile and aged women. Fertil Steril 34: 552-556

Nishizuka Y, Sakakura T (1969) Thymus and reproduction: sex-linked dysgenesis of the gonad after neonatal thymectomy in mice. Science 166: 753-755

Nishizuka Y, Sakakura T (1971) Ovarian dysgenesis induced by neonatal thymectomy in the mouse. Endocrinology 89: 886-893

Nishizuka Y (1982) A novel experimental system of organ-localized autoimmune diseases in the mouse. Acta Pathol Jpn 32: 211-222

Noda Y, Mori T, Takai I, Kohda H, Nishimura T (1981) A new method for preparation of pure zonae pellucidae in large quantities from porcine ovaries. J Reprod Immunol 3: 147-156

Oikawa T, Yanagimachi R (1975) Block of hamster fertilization by anti-ovary antibody. J Reprod Fertil 45: 487-494

Ownby CL, Shivers CA (1972) Antigens of the hamster ovary and effects of anti-ovary serum on eggs. Biol Reprod 6: 310-318

Sacco AG (1977) Antigenic cross-reactivity between human and pig zona pellucida. Biol Reprod 16: 164-173

Sacco AG, Moghissi KS (1979) Anti-zona pellucida activity in human sera. Fertil Steril 31: 503-506

Shahini SK, Padbidri JR, Rao SS (1972) Immunological studies with the reproductive organs, adrenals, and spleen of the female mouse. Int J Fertil 17: 161-165

Shivers CA, Dunbar BS (1977) Autoantibodies to zona pellucida: a possible cause for infertility in women. Science 197: 1082-1084

Taguchi O, Nishizuka Y (1980) Autoimmune oophoritis in thymectomized mice: T cell requirement in adoptive cell transfer. Clin Exp Immunol 42: 324-331

Taguchi O, Nishizuka Y, Sakakura T, Kojima A (1980) Autoimmune oophoritis in thymectomized mice: detection of circulating antibodies against oocytes. Clin Exp Immunol 40: 540-553

Takai I, Mori T, Noda Y, Nishimura T (1981) Heteroimmunization with isolated human ova. J Reprod Fertil 61: 19-24

Trounson AO, Shivers CA, McMaster R, Lopata A (1980) Inhibition of sperm binding and fertilization of human ova by antibody to porcine zona pellucida and human sera. Arch Androl 4: 29-36

Tsunoda Y, Chang MC (1976) Effect of anti-ovary antiserum on the fertilization of the mouse and hamster eggs in vivo and in vitro. Biol Reprod 14: 354-361

Tsunoda Y, Chang MC (1978) Effect of antisera against eggs and zonae pellucidae on fertilization and development of mouse eggs in vivo and in culture. J Reprod Fertil 54: 233-237

Tung KSK, Smith S, Matzner P, Kasai K, Oliver J, Feuchter F, Anderson RE (1987) Murine autoimmune oophoritis, epididymoorchitis, and gastritis induced by day 3 thymectomy: autoantibodies. Am J Pathol 126: 303-314

Wood DM, Liu C, Dunbar BS (1981) Effect of alloimmunization and heteroimmunization with zonae pellucidae on fertility in rabbits. Biol Reprod 25: 439-450

Yanagimachi R, Lopata A, Odom CB, Bronson RA, Mahi CA, Nicolson GL (1979) Retention of biologic characteristics of zona pellucida in highly concentrated salt solution: the use of salt-stored eggs for assessing the fertilizing capacity of spermatozoa. Fertil Steril 31: 562-574

Use of the Zona Pellucida as an Immunocontraceptive Target Antigen

A. G. Sacco and E. C. Yurewicz

Introduction

The zona pellucida, as a noncellular glycoprotein layer surrounding the mammalian oocyte and preimplantation embryo, occupies a rather unique position withir the female reproductive system for its consideration as a potential target fo immunocontraceptive purposes. The zona represents a structure through whicl sperm must pass in order to reach and fertilize the egg. Likewise, if an egg ha been fertilized, the zona pellucida then represents a structure which must b removed prior to the implantation of the developing embryo and initation of th pregnancy. Thus, in situations of either pre- or postfertilization the alteration of o interference with the normal functioning of the zona pellucida would likely resul in inhibition of certain critical reproductive events. Such disruption of norma reproductive function directed against the zona pellucida represents a potentia approach to fertility regulation.

The identification of the zona pellucida as a potential target antigen for contra ceptive vaccine development was made indirectly through studies originall designed to detect the presence of specific antigens associated with female ham ster and rabbit reproductive tissues (Ownby and Shivers 1972; Sacco and Shiver 1973 a, b). These studies, which utilized the Ouchterlony double diffusion tech nique and anti-ovary sera exhaustively absorbed with nonreproductive tissue, wer the first to demonstrate the existence of tissue-specific antigens within the ovar Further examination revealed that such ovary-specific antigens were associate with the zona pellucida (Sacco and Shivers 1973 c). Additional experimentatio (Ownby and Shivers 1972; Sacco and Shivers 1973 b) involving the treatment c ovulated zona-encased eggs from each of these species with their respectiv absorbed anti-ovary sera, indicated that such antibody treatment resulted in th formation of an antigen-antibody produced precipitation layer on the zona surfac (Fig. 1). This zona precipitate, which was manifested as an alteration in the ligh scattering properties of the zona, rendered the zona resistant to lytic digestion b proteolytic enzymes such as pronase and trypsin. These observations represente the first suggestion of the possible contraceptive potential of antibodies directe against the zona pellucida. Although earlier observations (Porter 1965; Porter et a 1970; Shahini et al. 1972) had indicated that antibodies to whole ovary coul reduce fertility, the precise ovarian target of these antisera, their ovarian specifi ity, and the mechanism of the observed fertility reduction remained unknown.

Since sperm utilize proteolytic acrosomal enzymes in traversing egg inves ments, the logical extension of these first observations (Ownby and Shivers 197

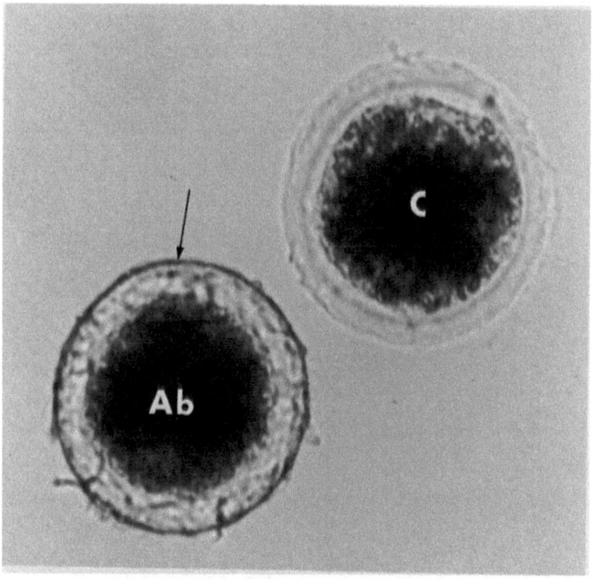

Fig. 1. Zona-encased pig eggs. Lower egg (*Ab*) has been treated with an antiserum produced against pig zonae pellucidae. Note dense precipitation layer (*arrow*) on outer surface of zona. Control egg (*C*) has been treated with preimmune serum and does not possess precipitation layer

using antibodies directed against zona was to examine what effect the presence of this antibody-produced zona precipitate had on sperm attachment, penetration, and fertilization. Such studies indicated that the presence of this precipitate on the zona surface totally or significantly inhibited sperm attachment and binding to the zona pellucida (Shivers et al. 1972; Shivers and Dudkiewicz 1974) thus unequivocally demonstrating the contraceptive potential of antibodies to the zona pellucida (Fig. 2).

Following these initial studies in the early 1970s considerable interest was generated in immunocontraceptive aspects of the zona pellucida. Ensuing investigations not only explored active and passive immunization approaches using the zona as immunogen but also involved studies on the species and tissue cross-reactivity of these zona antigens as well as intensive examination of the structure and macromolecular composition of the zona. As this work progressed, and knowledge of the composition of the zona pellucida expanded, immunocontraceptive studies gradually utilized antisera produced against "purer" and better defined zona pellucida components. Over the past 15 years production of antibodies to zona has progressed from using ovarian homogenates as the source of antigenic zona material to the use of zona-encased eggs, isolated intact zonae, isolated heat-solubilized zonae, and most recently the use of individual zona macromolecules.

The present chapter concentrates on use of the zona pellucida in immunocontraceptive investigations. However, some discussion of the macromolecular composition of the zona is first warranted to serve as an introduction to the biochemical and antigenic complexity of this structure. Such knowledge will be helpful in

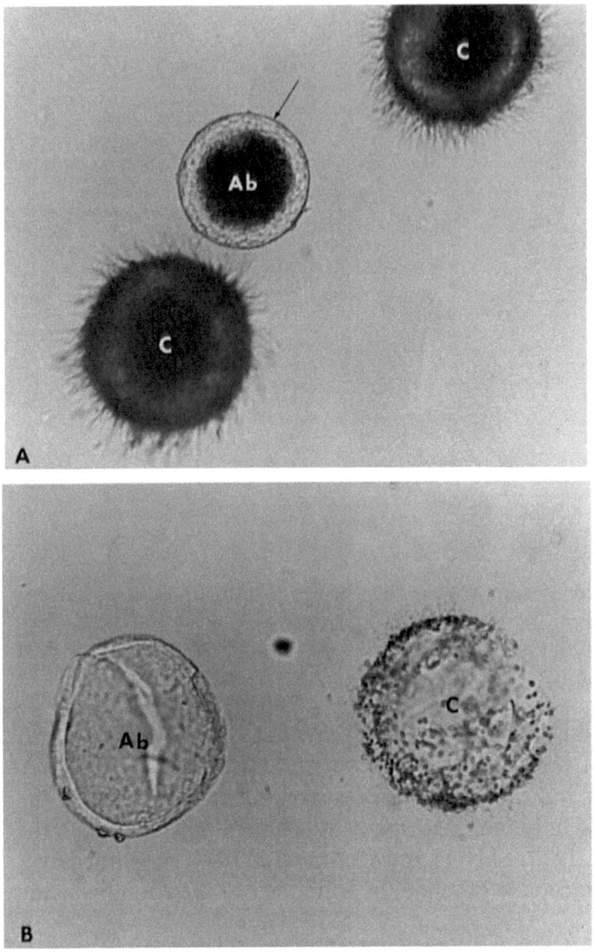

Fig. 2 A, B. Inhibition of sperm attachment by antibodies produced against the zona pellucida. **A** Zona-encased pig eggs treated with either zona antiserum *(Ab)* or preimmune serum *(C)*. Following serum treatment eggs were then exposed to boar sperm. Note how sperm readily attach to control eggs but are totally inhibited from attaching to antibody-treated egg. *Arrow* indicates precipitation layer on zona surface of antibody treated egg. **B** Human zonae treated with either antiserum to pig zonae *(Ab)* or preimmune serum *(C)*. Following serum treatment zonae were exposed to a human sperm preparation. Sperm readily attach to control treated zona but are inhibited in binding to antibody-treated zona. These results demonstrate the contraceptive potential of antibodies produced against pig zonae in the human system

understanding the nature and macromolecular content of the zona material used in early immunocontraceptive studies, as well as in comprehending the approaches and goals of the most recent studies and the direction of future strategies to be used. Knowledge of zona pellucida composition is restricted primarily to two species – the mouse and pig; this chapter will review only the pig zona since zonae from this species have been used in the vast majority of immunocontraceptive investigations. Information describing the composition of the mouse zona pellucida is available elsewhere in the literature (Bleil and Wassarman 1980 a; Shimizu et al. 1983) and in other chapters of the present volume.

Biochemistry of the Pig Zona Pellucida

Utilization of the porcine zona pellucida for immunocontraceptive studies has been greatly facilitated by recent advances in development of methodologies for purification of individual porcine zona pellucida macromolecules and characterization of their biochemical, immunological, and biological properties. Investigations employing one- and two-dimensional gel electrophoresis (Subramanian et al. 1981; Dunbar et al. 1981; Yurewicz et al. 1983; Hedrick and Wardrip 1986 a), conventional and high-performance liquid chromatography (Yurewicz et al. 1983, 1986, 1987), and chemical and enzymatic deglycosylations (Yurewicz et al. 1987; Sacco et al. 1986) have convincingly demonstrated that under nonreducing conditions the porcine zona pellucida is comprised of three major glycoproteins: one with an apparent molecular weight of 82000, referred to as 82-kd antigen, and two with an apparent molecular weight of 55000, referred to collectively as 55-kd antigen or individually as 55-kd antigen α- and β-glycoproteins. Under reducing conditions cleavage of disulfide bonds results in selective loss of 82-kd antigen and concomitant formation of products at 61-kd and 21-kd, indicating that 82-kd antigen is in fact a disulfide-linked dimer of 61-kd and 21-kd chains (Yurewicz et al. 1986; Hedrick and Wardrip 1986 a). Consequently, two-dimensional gel electrophoresis of nonreduced porcine zonae yields families of charge isomers at 82-kd and 55-kd whereas two-dimensional gel electrophoresis of reduced zonae yields families at 82-kd (ZP1), 61-kd (ZP2), 55-kd (ZP3), and 21-kd (ZP4) (Fig. 3). As discussed in greater detail below, electrophoretic separation of the 55-kd antigen α- and β-glycoproteins requires prior removal of carbohydrate.

Immunocontraceptive studies employing intact porcine zonae, heat-solubilized porcine zonae, and purified porcine zona 55-kd antigen have been reported in the literature (Sacco 1979; Sacco et al. 1981 a, 1983 a, 1987; Wood et al. 1981; Mahi-Brown et al. 1982) and will be discussed later in this chapter. In contrast, similar studies employing purified porcine zona 82-kd antigen as immunogen have yet to be described. Consequently, attention will be focused on recent advances in the purification and characterization of the 55-kd antigen (ZP3).

Two strategies have been employed for isolation of 55-kd antigen from porcine zonae pellucidae. The first involves electroelution of 55-kd antigen from two-dimensional gel electropherograms of separated zona components (Hedrick and Wardrip 1986 a). The second involves conventional column chromatography techniques for purification of 55-kd antigen from SDS (sodium dodecyl sulfate) dissociated nonreduced porcine zonae (Yurewicz et al. 1987). Two key aspects of this latter approach are of prime importance for maintenance of the antigenic integrity of the purified 55-kd macromolecule for its subsequent use in immunocontraceptive studies. First, purification is achieved in the absence of reagents or conditions which might be expected to result in cleavage of disulfide bonds. Second, renaturation of purified 55-kd antigen is accomplished by the extraction of SDS and its subsequent dialysis from a solution of 6-M guanidine hydrochloride.

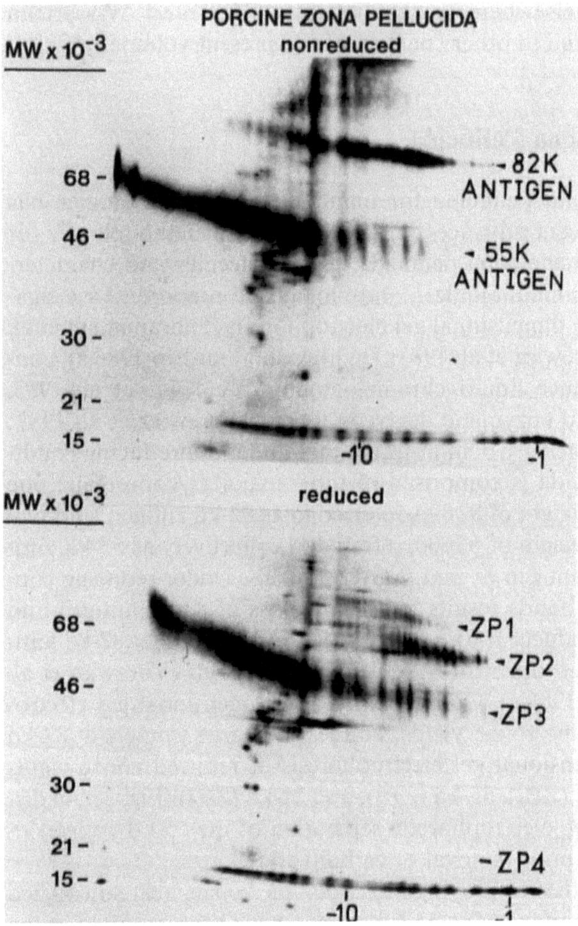

Fig. 3. Two-dimensional gel electrophoresis of nonreduced *(top)* and reduced *(bottom)* heat-solubilized porcine zonae pellucidae. Electrophoretic analysis was performed using isoelectric focusing in the first dimension (horizontal, acidic end to *left*) and SDS-PAGE in the second dimension (vertical, migration from *top* to *bottom*). Gels were calibrated with charge markers (carbamylated horse myoglobin) and molecular weight markers (reduced, carboxamidomethylated proteins). Proteins were visualized with Coomassie blue stain. The porcine zona pellucida 21-kd family of charge isomers (ZP4) is not readily apparent in two-dimensional gels of whole zonae

Purified 55-kd antigen migrates as a single diffuse band on SDS-PAGE (polycrylamide gel electrophoresis) and as a single family of acidic-charge isomers on two-dimensional gel electrophoresis under either reducing or nonreducing conditions (Fig. 4). Chemical analyses indicate the presence of carbohydrate (w/w, 30%) in addition to protein (Table 1). Interestingly, galactose and *N*-acetylglucosamine are present in elevated and approximately equimolar ratios suggestive of the presence of repeating *N*-acetyllactosamine units (Galβ1-4GlcNAc) termed lactosaminoglycans.

Contribution of lactosaminoglycan to the physicochemical and immunological

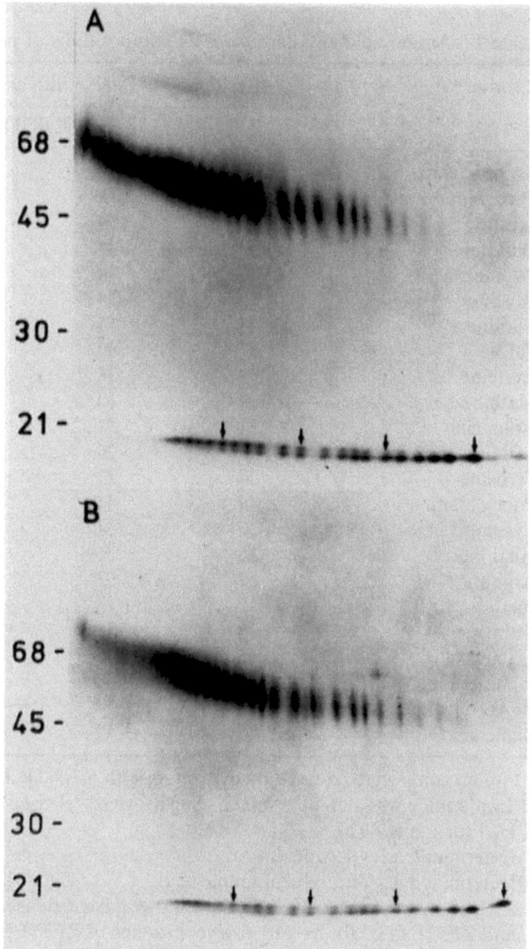

Fig. 4 A, B. Two-dimensional gel electrophoresis of purified 55-kd antigen of porcine zona pellucida. Electrophoresis was performed under either nonreducing **(A)** or reducing **(B)** conditions. Proteins were visualized with Coomassie blue stain. (From Yurewicz et al. 1987)

properties of ZP3 can be assessed by digestion with *Escherichia freundii* endo-β-galactosidase, an enzyme which specifically hydrolyzes galactosidic linkages of *N*-acetyllactosamine units in lactosaminoglycans. Incubation of 55-kd antigen with endo-β-galactosidase results in selective loss of approximately 33% of original saccharides (Table 1) with a concomitant dramatic alteration in electrophoretic behavior (Yurewicz et al. 1987). Both SDS-PAGE and two-dimensional gel electrophoresis (Fig. 5) resolve endo-β-galactosidase digested 55-kd antigen into two component glycoproteins with apparent molecular weights of 46000 (α-glycoprotein) and 42000 (β-glycoprotein). Compared to untreated 55-kd antigen, each exhibits markedly reduced charge heterogeneity and a shift to more neutral range of pI values. Enzymatic digestion of lactosaminoglycan is accompanied by an increase in immunoreactivity of the 55-kd antigen toward rabbit and squirrel mon-

Table 1. Amino acid and carbohydrate compositions of porcine 55-kd antigen (ZP3)

Residue [a]	Residues per 1000 amino acid residues [d]			
	ZP3	ZP3 (EβGD)	ZP3 (EβGDα)	ZP3 (EβGDβ)
Aspartic acid	82	84	77	86
Threonine	74	75	65	69
Serine	89	89	90	80
Glutamic acid	92	94	85	104
Proline	86	84	100	84
Glycine	55	54	55	55
Alanine	64	64	68	67
Valine	87	89	96	95
Cysteine [b]	39	38	45	35
Methionine	9	10	11	10
Isoleucine	31	31	35	28
Leucine	96	96	89	93
Tyrosine	24	22	30	9
Phenylalanine	44	44	35	43
Lysine	35	34	34	38
Histidine	35	36	31	40
Arginine	57	57	52	61
Fucose	12	13	12	14
Mannose	28	31	35	34
Galactose	86	41	31	38
N-Acetylglucosamine	94	59	53	59
N-Acetylgalactosamine	13	14	9	19
Sialic acid [c]	17	12	5	6

[a] Amino acids were released by hydrolysis with 6.7 *N* HCl for 24, 48, and 72 h at 110 °C in vacuo. Tryptophan was not determined. Sugars were released by hydrolysis with methanol −0.5 *N* HCl for 7 h at 80 °C.
[b] Determined as cysteic acid.
[c] Determined as *N*-acetylneuraminic acid.
[d] ZP3, 55-kd antigen; ZP3 (EβGD), endo-β-galactosidase digested 55-kd antigen; ZP3 (EβGDα) and ZP3 (EβGDβ), α- and β-glycoproteins of ZP3 (EβGD), respectively.

key anti-55-kd sera, suggesting that few if any major antigenic determinants were associated with the lactosaminoclycan removed by this procedure (Yurewicz et al. 1985, 1987).

High-performance liquid chromatography of endo-β-galactosidase digested 55-kd antigen resolves the α- and β-glycoproteins, thus permitting detailed studies of their comparative biochemical and immunological properties (Yurewicz et al. 1987). Biochemical investigations established that the α- and β-glycoproteins are structurally distinct entities. Chemical analyses (Table 1), amino-terminal sequence analyses, and peptide mapping experiments (Fig. 6) all support the hypothesis that the α- and β-glycoproteins possess unique primary structures (amino acid sequences) of their respective polypetide moieties.

Likewise, the purified 55-kd antigen α- and β-glycoproteins can also be readily distinguished on the basis of their immunological properties. Antisera were obtained from male rabbits immunized with either 55-kd antigen or purified α- and β-glycoproteins. The specificity of the resultant antibodies for homologous

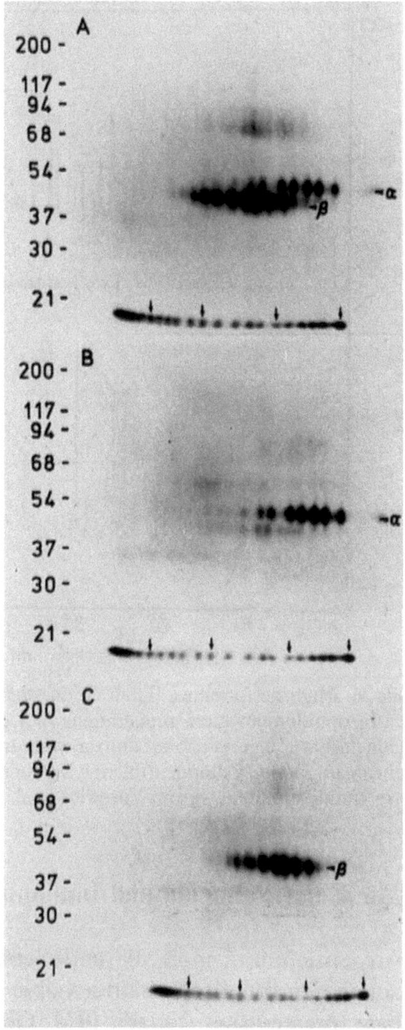

Fig. 5 A–C. Two-dimensional gel electrophoresis of endo-β-galactosidase digested 55-kd antigen **(A)** and purified α-**(B)** and β-glycoproteins **(C)**. The gels were stained with Coomassie blue. (From Yurewicz et al. 1987)

versus heterologous antigen was examined by radioimmunoassay (Fig. 7) and immunoblotting experiments (Fig. 8). The results clearly indicate that each antiserum preferentially recognizes its homologous antigen.

In summary, techniques are available for the purification of the major macromolecular components of the porcine zona pellucida, and significant advances have been made in recent years in understanding the composition of this structure. Concerted and ongoing investigations of the biochemical, immunological, and biological properties of the individual components of the zona should permit adequate evaluation of each as a potential target immunogen for antifertility vaccine development.

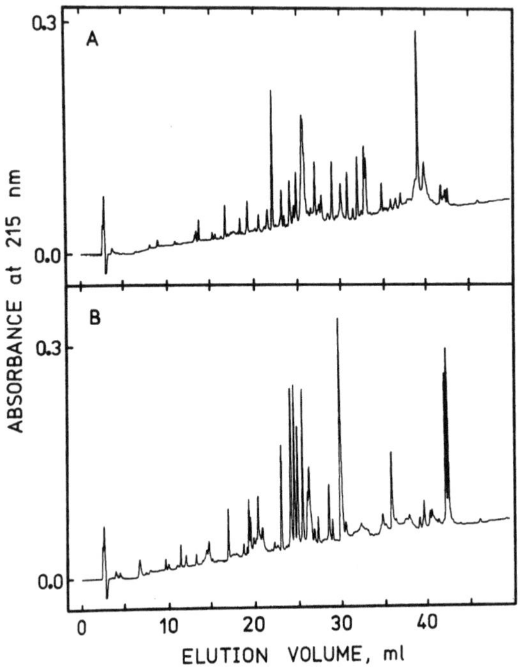

Fig. 6. High-performance liquid chromatography of tryptic peptides. Trypsin digests of reduced, carboxamidomethylated preparations of α-glycoprotein **(A)** and β-glycoprotein **(B)** of endo-β-galactosidase digested 55-kd antigen were fractionated by reverse-phase high-performance liquid chromatography. Column effluent was monitored at 215 nm. Major differences in peptide profiles are readily apparent. (From Yurewicz et al. 1987)

The Zona Pellucida and Immunocontraception

Any structure or molecule under consideration as an immuncontraceptive target antigen, in order to remain a viable candidate antigen, must satisfy a number of basic prerequisites (Celada 1974; Gillett 1977; Bahl and Muralidhar 1980; Sacco 1981; Anderson and Alexander 1983). How the zona pellucida meets the major prerequisites required for antigen consideration for immunocontraceptive purposes is outlined below.

The antigen must be specific to the reproductive system.

As indicated earlier, the zona pellucida, as a structure itself, is reproduction-specific. Over the years, careful analyses of its antigenic properties and tissue specificity by a variety of different laboratories and techniques have also convincingly demonstrated that the zona contains reproductive tissue-specific antigens as well. Such antigens are defined as those which are restricted to a particular reproductive tissue and are not present in any other tissue in the body. Techniques used to identify such antigens in the zona have included immunodiffusion and/or immunofluorescence (Ownby and Shivers 1972; Sacco and Shivers 1973 a, c; Sacco

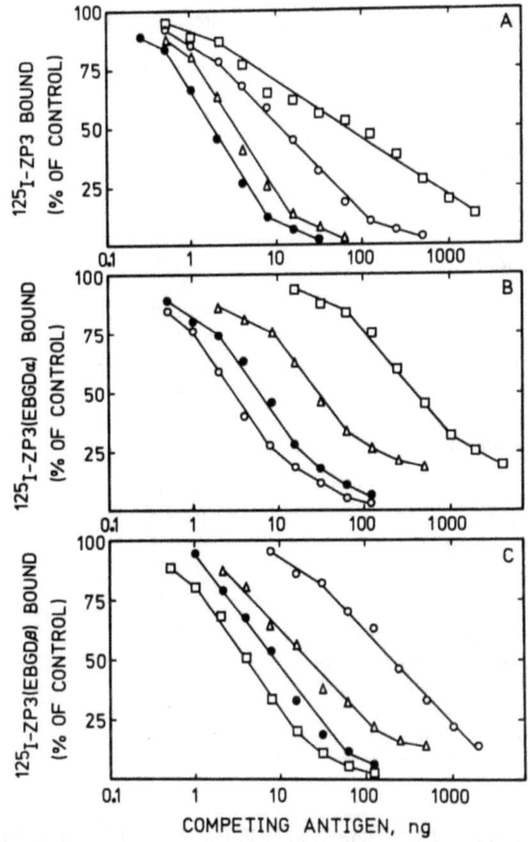

Fig. 7. Relative immunoreactivities of purified porcine zona pellucida antigens as analyzed by competitive binding radioimmunoassay. Antisera used includet: **A** anti-55-kd antigen; **B** anti-α-glycoprotein; **C** anti-β-glycoprotein. The radioinert competitors in each system were 55-kd antigen (-△-), α-glycoprotein (-○-), β-glycoprotein (-□-), and a reconstituted mixture w/w of protein (1:1), of α- plus β-glycoprotein (-●-). (From Yurewicz et al. 1987)

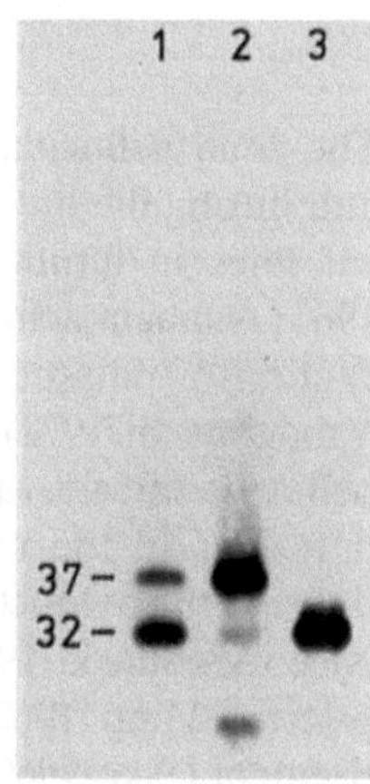

Fig. 8. Immunoblotting of deglycosylated α- and β-glycoproteins. Deglycosylated (trifluoromethanesulfonic acid treated) 55-kd antigen was run on an SDS-polyacrylamide gel to separate α- and β-polypeptides (37-kd and 32-kd, respectively). *Lane 1*, Coomassie blue staining; *lane 2*, immunostaining with anti-α-glycoprotein serum of western blot; *lane 3*, immunostaining with anti-β-glycoprotein serum of Western blot. (From Yurewicz et al. 1987)

Table 2. Levels of ZP3 immunoreactivity detected in squirrel monkey tissues and fluids by radioimmunoassay

Tissues	Percent ZP3 Immunoreactivity
Adrenal gland	0.0025
Adipose	0.0019
Brain	0.0011
Heart	0.0007
Kidney	0.0008
Liver	0.0009
Lung	0.0009
Ovary	0.0011
Oviduct	0.0012
Pancreas	0.0009
Serum	0.0025
Stomach	0.0006
Testis	0.0012
Thyorid	0.0005
Skeletal muscle	0.0004
Small intestine	0.0007
Spleen	0.0005
Uterus	0.0009
Human serum (male)	0.0001
Human serum (female)	0.0002
Human follicular fluid	0.0002
Pig serum	0.0001
ZP3 (1:10000 dilution)	96.1

1977 a, 1979; Sacco and Palm 1977; Jilek and Pavlok 1975; Gwatkin et al. 1979 a) organ distribution and clearance of radiolabeled zona antiserum in injected animals (Yanagimachi et al. 1976), and various radioimmunoassay (RIA) (Palm et al. 1979; Gerrity et al. 1981) and enzyme-linked immunoassay (ELISA) methods (Surneva-Nakova et al. 1984) (Table 2).

The functional role of the antigen should be critical to the normal reproductive process.

The zona pellucida likely has several functions, both in the fertilization process and during the preimplantation stages of early embryonic development. Two critical roles in fertilization include sperm recognition (species specificity) (Austin 1961; Yanagimachi 1972; Hanada and Chang 1972; Bedford 1977) and the prevention of polyspermy (Austin 1961; Braden et al. 1954; Piko 1969; Barros and Yanagimachi 1972). However, other functional roles also suggested for the zona pellucida include (a) the physical protection of both the oocyte (Glass 1970) while it resides in the ovary and of the embryo (Gwatkin 1963) during its journey through the oviduct, (b) the prevention of egg fusion (Mintz 1962) or of egg adherence (Modlinski 1970) to the oviduct, (c) the maintenance of a normal cleavage pattern (Mintz 1962), and (d) the osmotic regulation of the egg and embryo environment (Piko 1969).

With the recent elucidation of the macromolecular composition of the porcine zona pellucida, tentative functions have also been ascribed to its major macromolecules. The 82-kd and 61-kd glycoprotein families have demonstrated a substrate specificity for the sperm acrosomal enzyme acrosin and may be involved in sperm penetration of the zona (Brown and Cheng 1985; Dunbar et al. 1985; Urch et al. 1985). The 55-kd component has been shown to possess sperm receptor activity (Sacco et al. 1984; Berger et al. 1986) and may therefore play a role in the species-specificity and sperm-attachment aspects of the fertilization process.

Interference or inhibition of any of the above zona functions by antibodies directed to the zona could likely result in the disruption of a normal and critical reproductive process which in turn would lead to the subsequent inhibition of fertilization or development. Most evidence to date, however, indicates that one of the primary mechanisms whereby zona antibodies can inhibit fertility is by the prevention of normal sperm-egg interaction at the level of the zona pellucida. Zona antibodies appear to inhibit fertilization either by binding directly to sperm receptor molecules present in the zona or by obscuring these receptors by a steric hindrance mechanism. As indicated, such sperm receptor molecules have recently been identified and isolated from zonae of the mouse (Bleil and Wassarman 1980 b; Florman and Wassarman 1985; East et al. 1985) and pig (Sacco et al. 1984; Berger et al. 1986).

The antigen must be tissue specific but not necessarily species specific.

While the studies cited above have clearly shown the tissue specificity of certain zona antigens, early investigations indicated that zona antigens were shared among a number of different species. Antisera produced against zona of one species would cross-react in varying degrees of intensity with the zonae of several other species (Sacco 1981; Dunbar 1983) (Table 3). Of particular importance was the demonstration that antisera raised against zonae from several domestic species

Table 3. Species cross-reactions of zona antisera

Species zona antiserum		
Raised against	Cross-reacts with	Reference
Hamster	Mouse	Tsunoda and Chang (1976 a)
Rat	Mouse, hamster	Tsunoda and Chang (1976 b)
Mouse	Rat, hamster	Tsunoda and Chang (1976 c, 1977, 1978)
Hamster	Mouse, rhesus monkey, squirrel monkey	Gwatkin et al. (1977)
Human	Pig	Sacco (1977 b)
Marmoset	Pig, chimpanzee	Shivers et al. (1978)
Pig	Chimpanzee, marmoset	Shivers et al. (1978)
Human	Pig, marmoset, chimpanzee	Shivers et al. (1978)
Cow	Rabbit, rhesus monkey, marmoset, dog, human, hamster	Gwatkin et al. (1979), Gwatkin and Williams (1978)
Pig	Rabbit, squirrel monkey, human	Sacco et al. (1981 a)
Pig	Rabbit	Wood and Dunbar (1981)
Rabbit	Pig	Wood and Dunbar (1981)

(e.g.; pig, cow) cross-reacted with human zonae (Gwatkin et al. 1979 a; Gwatkin and Williams 1978; Sacco 1977 b; Shivers and Dunbar 1977). These studies thus identified species from which large numbers of zonae were readily obtainable for use in required biochemical, immunological, and immunocontraceptive studies. Since these zonae shared common antigens with human zonae, data derived from such investigations should also be readily applicable to the human system.

Antigen must be highly immunogenic.

The greater the immunogenicity of a particular antigen, the lesser the amount of antigen required to produce an adequate immune response. High immunogenicity is therefore a desirable attribute for an immunocontraceptive antigen since small quantities of antigen and/or few immunizations would be required to induce high-titered antiserum. Also, potent adjuvants or carrier molecules may not be required in the vaccine formulation. All studies utilizing the zona pellucida have demonstrated the potent immunogenicity of this glycoprotein material (Sacco 1981; Dunbar 1983). This was particularly apparent from early investigations which consistently obtained high-titered zona antibodies in response to immunization with ovarian homogenates and followed by exhaustive antiserum absorption. Significant titers of zona antibody were obtained despite the minimal amounts of zona protein present among the total ovary proteins used in these immunizations. A more recent study has shown that a minimum quantity of 1.65-3.3 μg zona protein injected into rabbits is required for a detectable immune response, and that only 40 μg is required to induce an antifertility effect (Dietl et al. 1982).

Antigen must be available in large quantities or be capable of being produced synthetically or by genetic-engineering methods.

The extremely small size and limited availability of mammalian zonae have for many years greatly restricted the number and types of studies which could be performed using zonae pellucidae material. The recent development of new zonae collection protocols (Oikawa 1978; Gwatkin et al. 1979b; 1980; Dunbar et al. 1980; Noda et al. 1981), however, has now made available adequate amounts of zonae material from certain species such as pig, cow, and rabbit. These collection protocols involve the passage of ovarian material or homogenates through a series of either nylon or metal screens of decreasing pore size to collect hundreds of thousands of zonae-encased oocytes. The oocytes are then passed through a tissue grinder to remove the zonae and the isolated zonae are collected on screens. Although there are several modifications of this basic protocol, all yield large quantities of isolated zonae material adequate for required biochemical, physiochemical, immunological, and immunocontraceptive studies. These latter investigations should lead to protocols for acquiring zona material by either peptide synthesis or genetic-engineering methods.

Adequate amounts of antibodies produced against the antigen must be capable of reaching the antigen in situ.

An immunocontraceptive target antigen must be localized within the reproductive system at a site readily accessible to humoral or cellular immunologic attack. The zona-encased oocyte residing in the developing follicle is bathed in follicular fluid, a substance very similar in composition to serum and previously demonstrated to

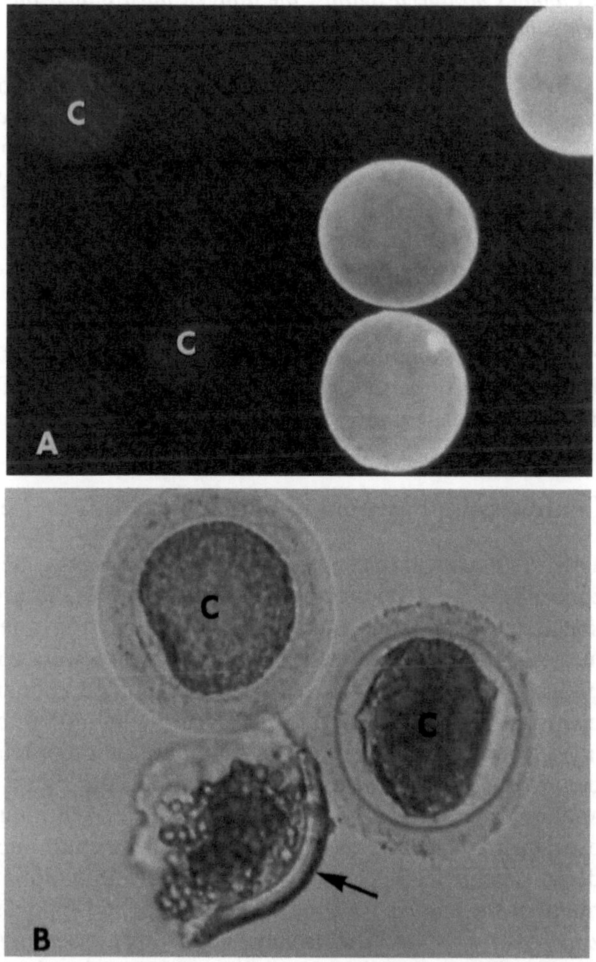

Fig. 9 A, B. In situ localization of zona antibodies. **A** Zona-encased eggs obtained from a mouse passively immunized with rabbit antiserum to mouse zona pellucida. Following passive immunization and superovulation protocol, reproductive tract of mouse was flushed to obtain eggs which were then treated with an anti-rabbit serum labeled with fluorescein isothiocyanate. Zonae of eggs fluoresce brightly, indicating prescence of anti-zona serum. Control eggs *(C)* were obtained from a nonimmunized mouse and do not fluoresce. **B** Zona-encased egg obtained by laparoscopy from a squirrel monkey actively immunized with the 55-kd antigen from pig zona pellucida. Note presence of antigen-antibody precipitate on zona surface *(arrow).* Control eggs *(C)* were obtained from a nonimmunized monkey and do not have precipitate on zona surface

contain immunoglobulin (Matousek et al. 1986; Kay et al. 1985; Hussein and Bourne 1984). Numerous investigations (Sacco 1981; Dunbar 1983) involving immunization of females with zona antisera have indicated that zona antibodies can indeed reach and combine with the zona pellucida in the follicle. Zona-encased eggs obtained from such immunized females have all possessed zona antibody on the zona surface, as established by microscopic or immunofluorescent procedures (Sacco 1979; Sacco et al. 1983a) (Fig. 9).

Antibodies produced against the antigen must inhibit the functional role of the antigen, which in turn results in the inhibition of fertility.

Many in vitro experiments which tested the effect of zona antisera on sperm-egg interaction and fertilization, as well as numerous passive and active immunization studies, have all indicated the potent contraceptive potential of antibodies produced against the zona pellucida (Sacco 1981; Dunbar 1983). These studies will be summarized in more detail below.

The specific antigen-antibody reaction must produce no deleterious physiological effect other than the inhibition of fertility.

How well the immunocontraceptive approach utilizing the zona pellucida satisfies this particular prerequisite is still uncertain due to conflicting and/or insufficient data. This question will be discussed in further detail later in this chapter.

Contraceptive Potential of Antibodies to Zona Pellucida

The first studies in the early to mid 1970s describing the contraceptive potential of zona antibodies generated tremendous interest in this area of research. A survey of the literature on zona immunocontraception from the mid to late 1970s yields numerous investigations, the majority of which were conducted in rodent species (mouse, rat, hamster), and which concentrated primarily upon three areas: (a) investigations evaluating the effect of zona antibodies on in vitro fertilization, (b) investigations involving the passive immunization of females with antisera to zona pellucida, and (c) investigations examining the tissue and species specificity of antibodies to zona. The zona antisera used were initially produced by utilizing ovarian homogenates as the immunizing material. Such antisera were then rendered specific to the zona by absorptions with various somatic tissues. However, many of the studies near the end of this period report using either isolated intact zonae or heat-solubilized isolated zonae for antiserum production. These antisera usually did not require absorbtion to be rendered specific to the zona. Active immunization with zonae material during the mid to late 1970s was performed primarily for the production of antisera for use in these in vitro and passive immunization projects. Very few studies were specifically undertaken at this time to evaluate the effect of active immunization with zona on fertility, and the few which were described involved very small numbers of animals.

Several excellent review articles are available which present detailed summaries and chronological lists of the many in vitro and passive immunization investigations of the zona performed during this period (Sacco 1981; Dunbar 1983; Shivers 1977; Aitken et al. 1984). The major findings and conclusions of these studies are the following:

1. The zona pellucida is a potent heteroimmunogen but a weak isoimmunogen.
2. The zona contains at least one tissue-specific antigen.
3. Antigens associated with the zona pellucida are not species specific.
4. Treatment of zonae with antibodies to zona causes a distinct antigen-antibody precipitation layer to form on the zona surface.
5. The presence of zona antibodies on the zona surface inhibits sperm attachment and fertilization. Zona shedding may also be inhibited.
6. Zona antibodies can reach and react with the zonae of oocytes contained in ovarian follicles.
7. Zona antibodies can inhibit sperm attachment to zona in both homologous (e.g., anti-mouse zona versus mouse zona) and heterologous (e.g., anti-pig zona versus monkey zona) situations.
8. Antibodies to zona pellucida possess contraceptive potential.

Selection of the Pig Zona as an Immunocontraceptive Target Antigen

As indicated above, initial immunocontraceptive zona studies involved the use of zona material from various rodent species. While data from these investigations were essential for demonstrating the contraceptive potential of zona antibodies, rodent species could not provide adequate quantities of zona material for the required large-scale biochemical and immunocontraceptive studies. Therefore, the major objective of investigations designed to examine the species cross-reactivity of zona pellucida antibodies was twofold: to identify a species (a) from which large numbers of zonae could be collected, and (b) which also shared common antigen(s) with the human zona pellucida. The availability of large numbers of zonae pellucidae was essential to provide sufficient quantities of zona material for the biochemical and immunocontraceptive testing required for the evaluation of any candidate zona target antigen. In addition, since the ultimate goal of such research is to develop a human contraceptive vaccine, data from experiments using zonae from a species sharing common antigen(s) with the human would likely also have direct application to the human system. The species which appeared best to fulfill these two requirements was the pig. Pig ovaries are readily available from commercial slaughterhouses, and it was clearly shown that human and pig zonae share common antigen(s) (Sacco 1977b). Consequently, beginning in the late 1970s the majority of efforts directed toward the development of a zona-based contraceptive vaccine have evolved around the pig zona.

Development of a Zona-Based Contraceptive Vaccine

Laboratories working toward the development of a zona-based contraceptive vaccine generally agree that if such a vaccine does eventually becomes a reality, it will likely be patterned according to the following parameters:

1. The immunization approach to be used will be that of active heteroimmunization of females.
2. The target zona antigen will be a component of the porcine zona pellucida.
3. The target antigen comprising the vaccine will consist of either (a) a biochemically purified pig zona macromolecule or fragment thereof which contains epitopes similar to a human zona macromolecule or (b) a synthetically produced or a genetically engineered zona immunogen which duplicates the composition of the relevant portion of the porcine zona macromolecule.

Regardless of final vaccine formulation this approach obviously requires rigorously characterized porcine zona macromolecules.

Several technical advances pertinent to continued progress in this research area occurred the late 1970s, concurrent with the identification of the porcine zona as the most viable zona contraceptive target antigen and the realization of the necessity for detailed biochemical data on zona pellucida macromolecular structure and composition. These included the development and application of new methodologies for (a) zona pellucida collection, (b) zona antibody and antigen detection, and (c) zona antigen separation and visualization.

Zona Pellucida Collection. The development of techniques for collection and isolation of large quantities of isolated zonae pellucidae were obviously foremost and were essential for any advancement of studies on the zona pellucida. Three essentially similar techniques are in use, and although developed primarily for collection of porcine zonae, they can also be applied to other species (cow, rabbit) as well. The first, reported by Oikawa (1978), involves the homogenization of ovaries using a commercial meat grinder and the isolation of oocytes by passage of the disrupted tissue through stainless steel mesh of decreasing pore size. The second method, initially reported by Dunbar et al. (1980) and modified by Wood et al. (1981), involves the use of a specially constructed instrument composed of ganged razor blades on wheels to disrupt ovarian follicles. The disrupted material is then passed through a series of nylon monofilament screens of decreasing pore size to collect isolated oocytes and zonae. The third technique, reported by Sacco (1981), uses the commercial meat grinder in combination with the nylon screens. Modifications of these basic protocols have also been reported and are in use by various other laboratories (Gwatkin et al. 1979b, 1980; Noda et al. 1981). These methods produce comparable yields of isolated zonae, and each can provide the large quantities of zona material required for biochemical and immunocontraceptive zona studies. However, it has recently been reported (Hedrick and Wardrip 1986b) that techniques which introduce the enzyme collagenase into the buffer system are likely to produce altered zona macromolecules, and such procedures may no longer be appropriate for routine zona collection purposes.

Zona Antibody and Antigen Detection. The introduction of sensitive RIA techniques (Palm et al. 1979) both for the detection of zona material, i.e., competitive binding assays (Subramanian et al. 1981; Palm et al. 1979), and for the detection and titration of zona antibodies, i.e., direct binding assays (Gerrity et al. 1981; Sacco et al. 1981b), has resolved a major shortcoming of early immunocontraceptive investigations involving the zona. Such RIA techniques as well as ELISA protocols adapted to the zona system (Surneva-Nakova et al. 1984; Fox et al. 1981; Gulyas et al. 1983a) have generally replaced the earlier, more tedious, and less sensitive methods for antibody detection and titration, such as zona precipitation, immunofluorescence, and inhibition of zona lysis and sperm binding (Sacco and Shivers 1978). The development of the competitive-binding RIA system was particularly essential for the successful purification of the 55-kd zona pellucida macromolecule by chromatographic techniques (Subramanian et al. 1981; Yurewicz et al. 1983).

Zona Antigen Separation and Visualization. As indicated above, the application to the zona system of biochemical and immunochemical techniques such as one- and two-dimensional (Figs. 3-5), the electrophoretic transfer (blotting) of proteins (Fig. 8), and two-dimensional immunoelectrophoresis has been instrumental in studies concerning the investigation of the macromolecular composition of the zona pellucida. Two-dimensional PAGE, particularly combined with sensitive silver staining techniques, has resulted in the resolution of the zona pellucida into its individual macromolecular components (Bleil and Wassarman 1980a; Subramanian et al. 1981; Dunbar et al. 1981; Yurewicz et al. 1983, 1986, 1987; Hedrick and Wardrip 1986a). Combined with electrophoretic blotting methods, two-dimensional PAGE represents a sensitive approach to visualize the reaction of different zona antisera with individual zona macromolecules (Yurewicz et al. 1987; Drell and Dunbar 1984). In combination with electroelution methods (Hedrick and Wardrip 1986a) it has allowed the isolation of individual zona macromolecules. Two-dimensional immunoelectrophoresis (Wood and Dunbar 1981; Popevanov and Surneva 1981; Sacco et al. 1983b) and monoclonal antibodies developed against zona proteins (Drell and Dunbar 1984; Sojima et al. 1981, 1984; Koyama et al. 1985) are other techniques recently applied to the zona systems to examine the antigenicity of zona macromolecules.

As a result of these new technologies being applied to the area of zona pellucida research significant advances have been made in recent years regarding the elucidation of zona pellucida structure and composition. The state of our present knowledge of the macromolecular composition of the pig zona was reviewed earlier in this chapter, and other reviews for both the pig and mouse zona are available elsewhere in this volume.

Active Immunization Studies

The vast majority of initial data demonstrating the contraceptive nature of zona antibodies was obtained from in vitro or passive immunization experiments conducted predominantly in homologous systems, i.e., the zona antibodies were produced against the zonae of the species to be tested. Obviously, consideration of this approach for purposes of human contraception is not currently feasible. Therefore, for the testing of a contraceptive vaccine with human application the efficacy of active heteroimmunization with zona in the inhibition of fertility had to be thoroughly explored utilizing zonae which are both antigenically similar to

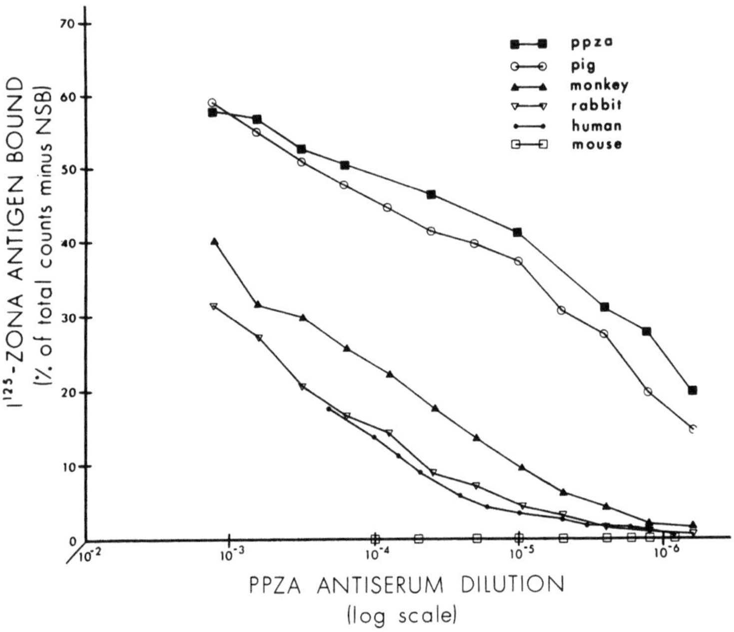

Fig. 10. Purified pig zona antigen antibody-binding profiles to ^{125}I-labeled PPZA and to ^{125}I labeled solubilized zona preparation from the pig, squirrel monkey, rabbit, human, and mous *PPZA*, purified pig zona antigen; this is comparable to the 55-kd antigen (ZP3). (From Sacco et a 1981 a)

human zonae, and which are available in adequate quantities. As indicated previ ously, the pig zona appears to fulfill these requirements.

In conducting such active immunization studies the most obvious test specie to use initially would be common laboratory rodents such as the mouse, rat, o hamster. However, a number of experiments have indicated that, while rodent an pig zonae do share similar cross-reacting antigens, the degree of cross-reactivity i not sufficient for antibodies to pig zona to exert a contraceptive effect in th rodent system (Sacco et al. 1981 a, c) (Fig. 10). Consequently, other species had t be considered when active immunization studies using pig zona as immunoge were undertaken in the early 1980s. Initial studies involved the active immuniza tions of dogs (Mahi-Brown et al. 1982, 1985; Shivers et al. 1981) and rabbit (Wood et al. 1981; Dietl et al. 1982; Skinner et al. 1984). More recent investiga tions have concentrated on nonhuman primates such as the squirrel monke (Sacco et al. 1983 a, 1987), cynomolgus monkey (Gulyas et al. 1983 b), marmos (Aitken et al. 1984), bonnet monkey (Bamezai et al. 1986), and baboon (Steven 1986). The decision to use nonhuman primates so early in active immunizatio studies involving the zona was therefore not one of choice but one of necessity.

Data from these studies demonstrated that active immunization with porcin zona results in a dramatic inhibitory effect upon fertility, thus once again empha sizing the contraceptive effectiveness of the zona-anti-zona approach to fertilit regulation. Furthermore, data from such active immunization studies also vividl emphasized that antibodies developed against porcine zonae do possess sufficient in situ cross-reactivity with the zonae material of the immunized species (i.e., the heterologous species) successfully to inhibit fertility. However, in some of these studies cycle irregularities and alterations in normal ovarian physiology were noted. Among the various species immunized it was with the rabbit - the species in which the most thorough studies have been conducted to date - that active immunization with heat-solubilized pig zona had the most severe effects upon ovarian function. Normal hormonal profiles were dramatically disrupted and the number of follicles in the ovaries of immunized animals was markedly reduced (Wood et al. 1981; Skinner et al. 1984). Various degrees of hormonal alterations were also observed in some of the studies using dogs and monkeys. However, other studies reported no adverse effects following active immunization with pig zona.

Thus, although producing encouraging contraceptive data, a major revelation of the active immunization experiments was that at least in some situations the observed infertility appeared to be induced primarily by zona antibodies inhibiting normal ovarian function rather than solely by fertility inhibition at the level of sperm-zona interaction. Such studies therefore reveal that zona antibodies can inhibit fertility by disrupting two major reproductive processes: (a) follicular development/differentiation and (b) sperm-egg interaction. Another emerging feature, however, as more active immunization data became available, was that effects on ovarian function in the nonhuman primate appeared to be considerably less severe and reversible compared to the adverse effects noted in the rabbit system. In fact, there is now some question as to whether the rabbit represents an appropriate species to consider for use in zona immunocontraceptive studies (Aitken et al. 1984).

Present Status of Zona Immunocontraceptive Research

While the question of ovarian dysfunction resulting from zona immunization is indeed an area for concern, these recent data indicating ovarian disturbances have not detracted from or discouraged the continued intensive investigation of zona macromolecules for contraceptive vaccine development. In fact, the observation that all active immunization studies to date have unequivocably demonstrated the potent contraceptive potential or antibodies to zona, has encouraged additional support for the continuation of such studies in this area.

The precise nature, extent, magnitude, and underlying mechanisms of the reported disturbances remain at present unclear. As noted, confusing and conflicting reports are found in the literature. For example, it has not yet been clarified whether the observed effect of zonae antibodies on ovarian function is attributable to the use of impure or contaminated zona preparations in these investigations. The failure to detect such hormonal and ovarian abnormalities in rabbits (Skinner and Dunbar 1984) in a follow-up study which used a purified protein rather than crude heat-solubilized zona material provides much optimism for the continuation of active immunization investigations using purified zona macromolecules. Although studies in the squirrel monkey (Sacco et al. 1983a, 1987) which used the purified 55-kd zona component as immunogen did detect hormonal irregularities, such disturbances were considerably less severe than those reported in the rabbit (Wood et al. 1981; Skinner et al. 1984), and they appear reversible. Also, both in studies on the squirrel monkey and in investigations conducted in marmosets (Aitken et al. 1984) there is some indication that such effects may be caused by stress or by the use of Freund's adjuvant rather than the immunogen. Studies in the bonnet monkey (Bamezai et al. 1986) which used intact porcine zonae (and, hence, impure zona preparations) as immunogen also indicated that the ovarian disturbances noted in the nonhuman primate were considerably less severe than those detected in the rabbit. Such findings from active immunization studies involving the nonhuman primate provide further support for the contention that the rabbit is not an appropriate species to use for immunocontraceptive studies using pig zona immunogen.

Since earlier investigations (Yanagimachi et al. 1976, Palm et al. 1979; Gerrity et al. 1981) have clearly demonstrated the tissue specificity of zona antigens by the failure of zona antibodies to cross-react with other body tissues, the observed effects at the level of the ovary must be caused by a specific reaction of zona antibodies with zona components. However, the exact cellular target(s) of these zona antibodies and their mechanism of action in producing such disturbances in normal ovarian function can at present only be speculated. The current hypothesis is that zona antibodies react with zona or zona precursor material at the site of zona synthesis and at approximately the time of follicle differentiation; this immune reaction and/or complex interferes with other cellular functions of the cell type responsible for zona synthesis, and this interference is manifested as a disturbance in normal hormonal profiles and folliculogenesis. Identification of the cell type involved (i.e., oocyte, granulosa cell, or both) and elucidation of the precise mechanism by which antibodies directed against zona components can produce other effects on cellular function are questions that remain to be resolved.

Present emphasis in the zona immunocontraceptive field is therefore directed toward identifying the smallest zona pellucida component that can elicit contraceptive zona antibodies which do not interfere with ovarian function. This zona component could be a purified intact zona macromolecule (e.g., 82-kd, 61-kd, 55-kd, or 21-kd component), a fragment of a zona macromolecule containing the pertinent epitopes, or a chemically or enzymatically modified (e.g., deglycosylated) fragment of a zona macromolecule.

Future Strategies and Directions in Zona Pellucida Immunocontraceptive Research

Two directions of research involving the zona pellucida are being pursued simultaneously. These include studies so designed: (a) to further our knowledge regarding the structure, composition, and antigenic nature of the zona and (b) to evaluate various zona pellucida immunogens in immunocontraceptive studies involving nonhuman primates.

Biochemical studies have concentrated on attemps to purify the major zona macromolecules either by chromatographic (Yurewicz et al. 1983, 1986, 1987) or electroelution procedures (Hedrick and Wardrip 1986a). The availability of even the small amounts of purified zona components obtained in this manner has permitted partial amino acid sequencing of the 55-kd component (Yurewicz et al. 1987). Since a major obstacle in studies on the zona pellucida has consistently been the difficulty in obtaining adequate amounts of material, recent emphasis has been on investigations directed ultimately toward obtaining zona proteins by either classical peptide-synthesis methods or by recombinant DNA technology. Since the immunogenicity of zona macromolecules and the epitopes producing antibodies with contraceptive properties may be dependent upon carbohydrate moieties, studies involving various deglycosylated zona macromolecules are also in progress (Yurewicz et al. 1987; Sacco et al. 1986). Knowledge concerning the immunogenicity of these modified zona macromolecules as well as the contraceptive efficacy of the antibodies they produce will be critical for efforts to obtain such components by peptide-synthesis or genetic-engineering methods. In addition, both polyclonal (Yurewicz et al. 1987; Sacco et al. 1983b) and monoclonal (Drell and Dunbar 1984; Isojima et al. 1981, 1984; Koyama et al. 1985) antibodies are being used to analyze the antigenic characteristics of different zona components. Collectively these various approaches should result in the identification and isolation of a number of new zona immunogens which can be tested for vaccine suitability in large scale immunocontraceptive studies.

Recently reported and presently ongoing immunocontraceptive investigations involve the active immunization of nonhuman primates. To date, only heat-solubilized total zona preparations or the purified 55-kd macromolecule have been used in these primate studies and with encouraging results. A total inhibition of fertility has been observed, and the effects on hormonal profiles and ovarian function seem relatively mild and reversible. Future studies will be directed toward the use of more refined zona components in conjunction with more tolerant adjuvant systems. As indicated above, the ultimate goal of such studies is to identify the smallest and least complex zona immunogen which will elicit antibodies producing their contraceptive effect only at the level of sperm-zona interaction while having no adverse effects on normal ovarian function. At present this appears to be a realistic and obtainable goal.

Among all the immunocontraceptive target antigens presently under consideration the zona pellucida is unique in that it does not represent a single molecular entity. Instead, it is a biochemically and antigenically complex structure composed of a multitude of potential target antigens, the majority of which can be considered for contraceptive vaccine purposes. Numerous investigations have convincingly shown that the zona-anti-zona approach can totally or significantly inhibit fertility prior to fertilization. While the use of current zona immunogens in active immunization studies has resulted in disturbances in ovarian function, the multitude and variations of zona components which remain to be tested provide much encouragement for this approach to fertility regulation. The demonstrated high effectiveness of this immunocontraceptive method offers much justification and support for continued intensive research efforts directed toward the ultimate goal of a zona-based contraceptive vaccine.

References

Aitken RJ, Richardson DW, Hulme M (1984) Immunological interference with the properties of the zona pellucida. In: Crighton DB (ed) Immunological aspects of reproduction in mammals. Butterworths, Boston, p 305

Anderson DJ, Alexander NJ (1983) A new look at antifertility vaccines. Fertil Steril 40: 557

Austin CR (1961) The mammalian egg. Thomas, Springfield

Bahl OP, Muralidhar K (1980) Current status of antifertility vaccines. In: Dhindsa D, Schumacher G (eds) Immunological aspects of infertility and fertility regulation. Elsevier/North-Holland, New York, p 225

Bamezai AK, Suman, Das C, Talwar GP (1986) Effect of immunization against porcine zona pellucida (PZP) on steroid hormone profiles and fertility in primates. J Reprod Immunol Suppl 85

Barros C, Yanagimachi R (1972) Polyspermy-preventing mechanism in the golden hamster eggs. J Exp Zool 180: 251

Bedford JM (1977) Sperm/egg interaction: the specificity of human spermatozoa. Anat Rec 188: 477

Berger T, Goldhawk DE, Wardrip NJ, Davis A, Hedrick JL (1986) Interaction of zona pellucida components with porcine sperm. J Cell Biol 103: 368a

Bleil JD, Wassarman PM (1980a) Structure and function of the zona pellucida: identification and characterization of the proteins of the mouse oocyte's zona pellucida. Dev Biol 76: 185

Bleil JD, Wassarman PM (1980b) Mammalian sperm-egg interaction: identification of a glycoprotein in mouse egg zonae pellucidae possessing receptor activity for sperm. Cell 20: 873

Braden AWH, Austin CR, David HA (1954) The reaction of the zona pellucida to sperm penetration. Aust J Biol 7: 391

Brown CR, Cheng WTK (1985) Limited proteolysis of the porcine zona pellucida by homologous sperm acrosin. J Reprod Fertil 74: 257

Celada F (1974) Concepts and methods in immunology applicable to the control of human fertility. In: Diczfalusy E (ed) Immunological approaches to fertility control. Bogtrykkeriet Forum, Copenhagen, p 419 (Karolinska symposia on research methods in reproductive endocrinology, 7th symposia)

Dietl J, Freye J, Mettler L (1982) Fertility inhibition using low-dose immunization with porcine zonae pellucidae. Am J Reprod Immunol 2: 153

Drell DW, Dunbar BS (1984) Monoclonal antibodies to rabbit and pig zonae pellucidae distinguish species-specific and shared antigenic determinants. Biol Reprod 30: 445

Dunbar BS (1983) Antibodies to zona pellucida antigens and their role in fertility. In: Wegmann TG, Gill TJ, Cumming CD, Nisbet-Brown E (eds) Immunology of reproduction. Oxford University Press, New York, p 507

Dunbar BS, Wardrip NJ, Hedrick JL (1980) Isolation, physicochemical properties, and the macromolecular of the zona pellucida from porcine oocytes. Biochemistry 19: 356

Dunbar BS, Liu C, Sammons DW (1981) Identification of three major proteins of porcine and rabbit zonae pellucidae by high resolution two-dimensional gel electrophoresis: comparison with serum, follicular fluid and ovarian cell proteins. Biol Reprod 24: 1111

Dunbar BS, Dudkiewicz AB, Bundman DS (1985) Proteolysis of specific porcine zona pellucida glycoproteins by boar acrosin. Biol Reprod 32: 619

East IJ, Gulyas BJ, Dean J (1985) Monoclonal antibodies to murine zona pellucida protein with sperm receptor activity: effects on fertilization and early development. Dev Biol 109: 268

Florman HM, Wassarman PM (1985) O-linked oligosaccharides of mouse egg ZP3 account for its sperm receptor activity. Cell 41: 313

Fox LL, Gengozian N, Shivers CA, Sieg P (1981) Marmoset response to immunization with porcine zonae pellucidae. J Med Primatol 10: 205

Gerrity M, Niu E, Dunbar BS (1981) A specific radioimmunoassay for evaluation of serum antibodies to zona pellucida antigens. J Reprod Immunol 3: 59

Gillett PG (1977) Immunologic control of fertility: search for a contraceptive vaccine. Clin Obstet Gynecol 20: 705

Glass LE (1970) Transmission of maternal proteins into oocytes. In: Raspe G (ed) Schering symposium on intrinsic and extrinsic factors in early mammalian development. Pergamon, Oxford p 29 (Advances in bioscience, vol 6)

Gulyas BJ, Yuan LC, Gwatkin RBL, Schmell ED (1983a) Response of monkeys to porcine zona pellucida as detected by a solid-phase radioimmunoassay. J Med Primatol 12: 331

Gulyas BJ, Gwatkin RBL, Yuan LC (1983b) Active immunization of cynomolgus monkeys (*Macaca fascicularis*) with porcine zona pellucida. Gamete Res 4: 299

Gwatkin RBL (1963) Studies on the zona pellucida of the mouse egg. J Reprod Fertil 6: 325

Gwatkin RBL, Williams DT (1978) Immunization of female rabbits with heat solubilized bovine zona: production of anti-zona antibody and inhibition of fertility. Gamete Res 1: 19

Gwatkin RBL, Williams DT, Carlo DJ (1977) Immunization of mice with heat solubilized hamster zonae: production of anti-zona antibody and inhibition of fertility. Fertil Steril 28: 871

Gwatkin RBL, Meyenhofer M, Williams DT (1979a) Localization of zona antigen in bovine ovary sections by fluorescent antibody. Gamete Res 2: 201

Gwatkin RBL, Williams DT, Meyenhofer M (1979b) Isolation of bovine zona pellucida from ovaries with collagenase: antigenic and sperm receptor properties. Gamete Res 2: 187

Gwatkin RBL, Anderson OF, Williams DT (1980) Large scale isolation of bovine and pig zonae pellucidae: chemical, immunological and receptor properties. Gamete Res 3: 217

Hanada A, Chang MC (1972) Penetration of zona-free eggs by spermatozoa of different species. Biol Reprod 6: 300

Hedrick JL, Wardrip NJ (1986a) Isolation of the zona pellucida and purification of its glycoprotein families from pig oocytes. Anal Biochem 157: 63

Hedrick JL, Wardrip NJ (1986b) Proteolysis by collagenase preparations alters the macromolecular composition of the porcine zona pellucida. Biol Reprod 35: 677

Hussein AM, Bourne FJ (1984) Immunoglobulin concentrations in pig follicular fluid. Int J Fertil 29: 54

Isojima S, Koyami K, Hasegawa A (1981) Production of monoclonal antibodies to zona pellucida from porcine oocytes. Acta Obstet Gynaecol Jpn 33: 1995

Isojima S, Koyama K, Hasegawa A, Tsunoda Y, Hanada A (1984) Monoclonal antibodies to porcine zona pellucida antigens and their inhibitory effects on fertilization. J Reprod Immunol 6: 77

Jilek F, Pavlok A (1975) Antibodies against mouse ovaries and their effect on fertilization in vitro and in vivo in the mouse. J Reprod Fertil 42: 377

Kay DJ, Boettcher B, Yovich JL, Stanger JD (1985) Antispermatozoal antibodies in human follicular fluid. Am J Reprod Immunol Microbiol 7: 113

Koyama K, Hasegawa A, Tsuji Y, Isojima S (1985) Production and characterization of monoclonal antibodies to cross-reactive antigens of human and porcine zonae pellucidae. J Reprod Immunol 7: 187

Mahi-Brown CA, Huang TTF Jr, Yanagimachi R (1982) Infertility in bitches induced by active immunization with porcine zonae pellucidae. J Exp Zool 222: 89

Mahi-Brown CA, Yanagimachi R, Hoffman JC, Huang TTF (1985) Fertility control in the bitch by active immunization with porcine zona pellucida: use of different adjuvants and patterns of estradiol and progesterone levels in estrous cycles. Biol Reprod 32: 761

Matousek J, Hojny J, Janik A (1986) Serum allotypes in ovarian follicular fluids of pigs. Anim Genet 17: 169

Mintz B (1962) Experimental study of the developing mammalian egg. Removal of the zona pellucida. Science 138: 594

Modlinski JA (1970) The role of zona pellucida in the development of mouse eggs in vivo. J Embryol Exp Morphol 23: 539

Noda Y, Mori T, Takai I, Kohda H, Nishimura T (1981) A new method for preparation of pure zonae pellucidae in large quantities from porcine ovaries. J Reprod Immunol 3: 147

Oikawa T (1978) A simple method for the isolation of a large number of ova from pig ovaries. Gamete Res 1: 265

Ownby CL, Shivers CA (1972) Antigens of the hamster ovary and effects of antiovary serum on eggs. Biol Reprod 6: 310

Palm VS, Sacco AG, Syner FN, Subramanian MG (1979) Tissue specificity of porcine zona pellucida antigen(s) tested by radioimmunoassay. Biol Reprod 21: 709

Piko L (1969) Gamete structure and sperm entry in mammals. In: Metz CB, Monroy A (eds) Fertilization, vol 2. Academic, New York, p 325

Popevanov RP, Surneva TN (1981) Immunoelectrophoretic studies on the antigens of porcine zona pellucida. C R Acad Bulg Sci 34: 741

Porter CW (1965) Ovarian antibodies in female guinea pigs. Int J Fertil 10: 257
Porter CW, Highfill D, Winovich R (1970) Guinea pig ovary and testis: demonstration of common gonad specific antigens in the ovary and testis. Int J Fertil 15: 171
Sacco AG (1977a) Antigenicity of the human zona pellucida. Biol Reprod 16: 158
Sacco AG (1977b) Antigenic cross-reactivity between human and pig zona pellucida. Biol Reprod 16: 164
Sacco AG (1979) Inhibition of fertility in mice by passive immunization with antibodies to isolated zonae pellucidae. J Reprod Fertil 56: 533
Sacco AG (1981) Immunocontraception: consideration of the zona pellucida as a target antigen. In: Wynn RM (ed) Obstetrics and gynecology annual, vol 10. Appleton-Century-Crofts, New York, p 1
Sacco AG, Palm VS (1977) Heteroimmunization with isolated pig zonae pellucidae. J Reprod Fertil 51: 165
Sacco AG, Shivers CA (1973a) Antigens of the rabbit ovary, oviduct and uterus. J Reprod Fertil 32: 403
Sacco AG, Shivers CA (1973b) Effects of reproductive tissue-specific antisera on rabbit eggs. Biol Reprod 8: 481
Sacco AG, Shivers CA (1973c) Localization of tissue-specific antigens in the rabbit ovary, oviduct and uterus by the fluorescent antibody technique. J Reprod Fertil 32: 415
Sacco AG, Shivers CA (1978) Immunologic inhibition of development. In: Daniel IC (ed) Methods in mammalian reproduction. Academic, New York, p 203
Sacco AG, Yurewicz EC, Subramanian MG, DeMayo FJ (1981a) Zona pellucida composition: species cross reactivity and contraceptive potential of antiserum to a purified pig zona antigen (PPZA). Biol Reprod 25: 997
Sacco AG, Subramanian MG, Yurewicz EC (1981b) Application of a radioimmunoassay (RIA) for monitoring immune response to porcine zonae pellucidae. Proc Soc Exp Biol Med 167: 318
Sacco AG, Subramanian MG, Yurewicz EC (1981c) Active immunization of mice with porcine zonae pellucidae: immune response and effect on fertility. J Exp Zool 218: 405
Sacco AG, Subramanian MG, Yurewicz EC, DeMayo FJ, Dukelow WR (1983a) Heteroimmunization of squirrel monkeys (*Saimiri sciureus*) with a purified porcine zona antigen (PPZA): immune response and biologic activity of antiserum. Fertil Steril 39: 350
Sacco AG, Yurewicz EC, Zhang S (1983b) Immunoelectrophoretic analysis of the porcine zona pellucida. J Reprod Fertil 68: 21
Sacco AG, Subramanian MG, Yurewicz EC (1984) Association of sperm receptor activity with a purified pig zona antigen (PPZA). J Reprod Immunol 6: 89
Sacco AG, Yurewicz EC, Subramanian MG (1986) Carbohydrate influences the immunogenic and antigenic characteristics of the ZP3 macromolecule (Mr 55000) of the pig zona pellucida. J Reprod Fertil 76: 575
Sacco AG, Pierce DL, Subramanian MG, Yurewicz EC, Dukelow WR (1987) Ovaries remain functional in squirrel monkeys (*Saimiri sciureus*) immunized with porcine zona pellucida 55 kd macromolecule. Biol Reprod 36: 481
Shahini SK, Padbidri JR, Rao SS (1972) Immunological studies with the reproductive organs adrenals, and spleen of the female mouse. Int J Fertil 17: 161
Shimizu S, Tsuji M, Dean J (1983) In vitro biosynthesis of three sulfated glycoproteins of murine zonae pellucidae by oocytes grown in follicular culture. J Biol Chem 258: 5858
Shivers CA (1977) The zona pellucida as a possible target in immunocontraception. In: Boettcher B (ed) Immunological influence on human fertility. Academic, New York, p 13
Shivers CA, Dudkiewicz AB (1974) Inhibition of fertilization with specific antibodies. In: Coutinho EM, Fuchs F (eds) Physiology and genetics of reproduction, part B. Plenum, New York p 81
Shivers CA, Dunbar BS (1977) Autoantibodies to zona pellucida: a possible cause for infertility in women. Science 197: 1187
Shivers CA, Dudkiewicz AB, Franklin LE, Fussell EN (1972) Inhibition of sperm-egg interaction by specific antibody. Science 178: 1211
Shivers CA, Gengozian N, Franklin S, McLaughlin CL (1978) Antigenic cross-reactivity between human and marmoset zonae pellucidae, a potential target for immunocontraception. J Med Primatol 7: 242

Shivers CA, Sieg PM, Kitchen H (1981) Pregnancy prevention in the dog: potential for an immunological approach. J Am Anim Hosp Assoc 17: 823
Skinner SM, Dunbar BS (1984) Comparsion of immunization using a purified zona pellucida (ZP) protein on ovarian follicular development. Biol Reprod Suppl [1] 30: 73
Skinner SM, Mills T, Kirchick HJ, Dunbar BS (1984) Immunization with zona pellucida proteins results in abnormal ovarian follicular differentiation and inhibition of gonadotropin-induced steroid secretion. Endocrinology 115: 2418
Stevens VC (1986) State of the art in immunological fertility control. J Immunol Immunopharmacol 6: 118
Subramanian MG, Yurewicz EC, Sacco AG (1981) Specific radioimmunoassay for the detection of a purified porcine zona pellucida antigen (PPZA). Biol Reprod 24: 933
Surneva-Nakova TN, Nakov LS, Kyurkchiev SD, Gancheva SD, Yalumov OI (1984) Study on tissue specificity of pig zona pellucida by enzyme-linked immunosorbent assay (ELISA). C R Acad Bulg Sci 37: 975
Tsunoda Y, Chang MC (1976a) In vivo and in vitro fertilization of hamster rat and mouse eggs after treatment with anti-hamster ovary antiserum. J Exp Zool 195: 409
Tsunoda Y, Chang MC (1976b) Effect of anti-rat ovary antiserum on the fertilization of rat, mouse and hamster eggs in vivo and vitro. Biol Reprod 14: 354
Tsunoda Y, Chang MC (1976c) The effect of passive immunization with hetero- and iso-immune anti-ovary antiserum on the fertilization of mouse, rat, and hamster eggs. Biol Reprod 15: 361
Tsunoda Y, Chang MC (1977) Further studies of antisera on the fertilization of mouse, rat, and hamster eggs in vivo and in vitro. Int J Fertil 22: 129
Tsunoda Y, Chang MC (1978) Effects of antisera on fertilization of mouse, rat and hamster eggs. Biol Reprod 18: 468
Urch UA, Wardrip NJ, Hedrick JL (1985) Limited and specific proteolysis of the zona pellucida by acrosin. J Exp Zool 233: 479
Wood DM, Dunbar BS (1981) Direct detection of two cross-reactive antigens between porcine and rabbit zonae pellucidae by radioimmunoassay and immunoelectrophoresis. J Exp Zool 217: 423
Wood DM, Liu C, Dunbar BS (1981) Effect of alloimmunization and heteroimmunization with zonae pellucidae on fertility in rabbits. Biol Reprod 25: 439
Yanagimachi R (1972) Penetration of guinea-pig spermatozoa into hamster eggs in vitro. J Reprod Fertil 28: 477
Yanagimachi R, Winkelhake JL, Nicolson GL (1976) Immunological block to mammalian fertilization: survival and organ distribution of immunoglobulin which inhibits fertilization in vivo. Proc Natl Acad Sci USA 73: 2405
Yurewicz EC, Sacco AG, Subramanian MG (1983) Isolation and preliminary characterization of a purified pig zona antigen (PPZA) from porcine oocytes. Biol Reprod 29: 511
Yurewicz EC, Sacco AG, Subramanian MG (1985) Role of lactosaminoglycan in antigenicity of the major glycoprotein family of porcine oocyte zona pellucida. In: Davidson EA, Williams IC, DiFerrante NM (eds) Glycoconjugates. Proceedings of the 8th International Symposium, vol 2. Praeger, New York, p 589
Yurewicz EC, Sacco AG, Subramanian MG (1986) Pathways to immunocontraception: biochemical and immunological properties of glycoprotein antigens of the porcine zona pellucida. In: Hedrick JL (ed) The molecular and cellular biology of fertilization. Plenum, New York, 94-123
Yurewicz EC, Sacco AG, Subramanian MG (1987) Structural characterization of the Mr=55000 antigen (ZP3) of porcine oocyte zona pellucida. J Biol Chem 262, 564-571

Subject Index

MIX
Papier aus verantwortungsvollen Quellen
Paper from responsible sources
FSC® C105338

If you have any concerns about our products,
you can contact us on
ProductSafety@springernature.com

In case Publisher is established outside the EU,
the EU authorized representative is:
Springer Nature Customer Service Center GmbH
Europaplatz 3, 69115 Heidelberg, Germany

Printed by Libri Plureos GmbH
in Hamburg, Germany